NASA Technical Memorandum 4255

Mercury Orbiter: Report of the Science Working Team

National Aeronautics and
Space Administration

Office of Management

Scientific and Technical
Information Division

1991

The Mercury Orbiter Science Working Team

John W. Belcher (Chairperson)	Massachusetts Institute of Technology
James A. Slavin (Study Scientist)	NASA/Goddard Space Flight Center
Thomas P. Armstrong (Program Scientist)	NASA Headquarters
Robert W. Farquhar (Program Manager)	NASA Headquarters
Syun I. Akasofu	University of Alaska
Daniel N. Baker	NASA/Goddard Space Flight Center
Cynthia A. Cattell	University of California at Berkeley
Andrew F. Cheng	JHU/Applied Physics Laboratory
Edward L. Chupp	University of New Hampshire
Pamela E. Clark	Caltech/Jet Propulsion Laboratory
Merton E. Davies	RAND Corporation
Edward W. Hones	Los Alamos National Laboratory
William S. Kurth	University of Iowa
J. K. Maezawa	Nagoya University (Japan)
Franco Mariani	Universita di Roma (Italy)
Eckart Marsch	Max-Planck-Institut fur Aeronomie (FRG)
George K. Parks	University of Washington
Edward G. Shelley	Lockheed Palo Alto Research Laboratory
George L. Siscoe	University of California at Los Angeles
Edward J. Smith	Caltech/Jet Propulsion Laboratory
Robert G. Strom	University of Arizona
J. I. Trombka	NASA/Goddard Space Flight Center
Donald J. Williams	JHU/Applied Physics Laboratory
Chen-wan Yen	Caltech/Jet Propulsion Laboratory

Executive Summary

The Mariner-10 flybys of Mercury in 1974 and 1975 resulted in the discovery of a planetary magnetic field and an active magnetosphere similar to that of Earth. Based upon the small size of the planet, Mercury's interior was expected to have cooled and solidified long ago. The presence of an intrinsic magnetic field, however, implied an internal dynamo in a fluid core, posing fundamental, unresolved issues concerning the origin, composition, and thermal history of Mercury. The Mariner-10 spacecraft also detected intense particle bursts and magnetic field disturbances, indicating that magnetospheric substorms occur at Mercury. The Mariner-10 images revealed a number of surface features unique to Mercury, including large-scale thrust faults apparently associated with crustal compression as the planet cooled and contracted. Follow-on missions to Mercury were studied in the late 1970's, but deferred because of perceived difficulties in spacecraft propulsion and thermal engineering requirements as they were understood at the time.

Within the past few years, it has become apparent that a moderate-cost mission to Mercury can provide the particles and fields measurements and planetological observations necessary to yield major advances in our understanding of Mercury and its magnetosphere. Mercury Orbiter (MeO) as described in this report is such a mission. It involves dual, spin-stabilized spacecraft launched by a single Titan-IV Centaur vehicle, a 4-5 year gravity-assist trajectory, and a nominal one Earth-year-duration mission at Mercury.

This report presents the results of the Mercury Orbiter Science Working Team (MeO SWT) which held three workshops in 1988/1989 under the auspices of the Space Physics and Planetary Exploration Divisions of NASA Headquarters. Spacecraft engineering and mission design studies at the Jet Propulsion Laboratory were conducted in parallel with this effort and are detailed elsewhere. The findings of the engineering study, summarized in this report, indicate that spin-stabilized spacecraft carrying comprehensive particles and fields experiments and key planetology instruments in highly elliptical orbits can survive and function in Mercury orbit without costly sun-shields and active cooling systems.

The magnetospheric and planetary physics rationale for a Mercury orbiter mission has been reported upon previously in the Report of the Terrestrial Bodies Science Working Group (JPL, 1977); Strategy for Exploration of the Inner Planets: 1977-1987 (NAS, 1978); Mercury Polar Orbiter (ESA, 1985); An Implementation Plan for Priorities in Solar-System Space Physics (NAS, 1985); and Space Science in the Twenty-First Century (NAS, 1988). The MeO SWT has refined and extended these previously identified science objectives and developed a strawman payload and mission plan which is responsive to the technical constraints placed on the spacecraft by Mercury's thermal environment and MeO's propulsive requirements. The primary space physics science objectives for MeO are: 1) to map in three dimensions the magnetic structure and plasma environment of this "miniature" magnetosphere; 2) to study in detail the principal physical processes taking place during Hermean magnetospheric substorms with an emphasis on differences from Earth due to Mercury's lack of a highly conducting ionosphere; 3) to assess the role of interplanetary conditions in determining the rate at which the Hermean magnetosphere draws energy from the solar wind and the manner in which it is later dissipated; 4) to investigate heliospheric structure and dynamics inside of 0.5 AU; and 5) to utilize the proximity of Mercury to the Sun to achieve fundamental solar physics objectives by measuring neutrons and charged particles emanating from flare regions. The primary planetology science objectives for MeO are: 1) to complete the global surface mapping initiated by Mariner-10; 2) to obtain global geochemical terrain maps of the occurrence of such elements as Fe, Th, K, Ti, Al, Mg, and Si; 3) to measure the intrinsic magnetic field in sufficient detail to allow for the detection of magnetic anomalies; and 4) to map Mercury's gravitational field and associated anomalies.

The MeO SWT has identified a ten-instrument strawman payload to meet these science objectives: magnetometer, electric field analyzer, plasma wave analyzer, energetic particle detector, fast plasma analyzer, ion composition analyzer, solar wind plasma analyzer, solar neutron detector, line-scan imager, and gamma/x-ray spectrometer. All of these instruments are based upon mature technologies and should require minimal development to meet the requirements of the MeO mission.

The MeO SWT strongly endorses the mission plan developed by the JPL study team. The single launch vehicle, dual spacecraft baseline meets the fundamental magnetospheric science requirements for simultaneous multipoint measurements and provides critical redundancy in the event of a spacecraft failure. The coordinated orbit scenarios for the two spacecraft will provide unique particles and fields measurements which are unobtainable at other planets due to the constraints of orbital mechanics and the large dimensions of other magnetospheres relative to their planetary bodies. In conjunction with the Earth-orbiting ISTP and CLUSTER missions to be flown in the 1990s, the Mercury Orbiter Mission will provide the essential data necessary to formulate the next generation of theories and models for terrestrial-type magnetospheric structure and dynamics. This mission will also return critical measurements necessary for the understanding of not just the surface history and internal structure of Mercury, but the formation and chemical differentiation of the Solar System as a whole.

TABLE OF CONTENTS

		Page
Science Working Team Membership		iii
Executive Summary		v
I.	Introduction	1
II.	Mercury Science Objectives	3
	2.1 Magnetospheric Physics	3
	2.2 Planetology	13
III.	Heliospheric and Solar Science Objectives	27
	3.1 Physics of the Inner Heliosphere	27
	3.2 Solar Flares and Solar Energetic Particles	36
IV.	Mercury Orbiter Mission	47
	4.1 Strawman Instruments	47
	4.1.1 Plasma and Energetic Particle Measurements	47
	4.1.2 Electric and Magnetic Field Instruments	52
	4.1.3 Surface Imager	60
	4.1.4 The Geochemistry Experiment	66
	4.1.5 X-Ray, Gamma Ray and Neutron Instrumentation for Solar Physics	74
	4.2 Spacecraft Design	77
	4.3 Mission Design	93
V.	Fundamental Problems in Planetary Science Addressed by MeO	109
	5.1 Magnetospheric Physics Summary	109
	5.2 Planetary Physics Summary	120
VI.	References	127
	Appendix A: MeO Design Team	

I. INTRODUCTION

The Mercury Orbiter (MeO) Science Working Team (SWT) was jointly appointed by the Space Physics (Code SS) and Planetary Exploration (Code SL) Divisions at NASA Headquarters. The science working team was international in composition with scientists from the U.S., Europe and Japan participating in the deliberations. Its charter was to develop a science rationale and mission scenario for a Mercury Orbiter Mission whose primary science objective would be the in-depth study of this body's very active magnetosphere. Secondary objectives were to be the investigation of the interplanetary medium inside of 1 AU, solar energetic particle acceleration, and the surface and interior of the planet itself. All of these disciplines were represented in the membership of the science working team appointed by NASA Headquarters.

The MeO SWT conducted three workshops in 1988-1989 where many different mixes of science and possible mission plans were discussed. All of these activities were supported by a JPL Mission Design Team led by Dr. C.-W. Yen, who conducted engineering feasibility studies in response to science requirements levied by the science working team. The findings and recommendations of the MeO SWT are detailed in this NASA Technical Memorandum and a mission design report (JPL D-7443). As will be presented in the chapters to follow, it is the conclusion of the MeO SWT that a Mercury Orbiter Mission is of the utmost scientific importance to not only the magnetospheric and planetary physics communities, but also to solar and heliospheric physics. Furthermore, the engineering studies conducted at JPL indicate that the mission science requirements can be met with existing spaceflight technology and within the scope of a NASA Office of Space Science and Applications "moderate" class mission. As presented in the sections to follow, it is our conclusion that the Mercury Orbiter Mission should be the subject of a formal phase-A study in preparation for launch opportunities in the late 1990s.

II. MERCURY SCIENCE OBJECTIVES

2.1 MAGNETOSPHERIC PHYSICS

The solar wind moves radially outward from the Sun at speeds of hundreds of kilometers per second. This collisionless plasma is composed predominantly of ionized hydrogen, H^+, and is threaded by magnetic field lines from the Sun. How each planetary body acquires and releases solar wind energy depends on the conditions in the upstream solar wind and the planet's intrinsic magnetic field and atmosphere. The electrically conducting solar wind does not easily penetrate a planet's magnetic field. Planets with intense magnetic fields, such as the Earth, do not encounter the solar wind directly. Instead, each forms a blunt bullet-shaped magnetic cavity, called a magnetosphere, in which the planet's intrinsic magnetic field is confined by the solar wind. On the sunward side, the planetary field is compressed and on the nightside, the field lines are stretched out into a long magnetotail. In such cases, the ultimate resistance to the solar wind's flow comes from the planet itself with the normal stress of the solar wind being transmitted downward by the magnetic field. The tangential drag on the magnetosphere is mediated by field-aligned electric currents (FACs) that flow from the magnetopause down to the planet. In order for these currents to be more than transient, the planet must have an ionosphere or a moderately conducting surface to close the circuit.

Measurements taken by instruments on the Mariner-10 spacecraft revealed that Mercury has an intrinsic magnetic field strong enough to stand off the solar wind at distances of ~1.5-2.0 Mercury radii (1 R_M = 2439 km) above

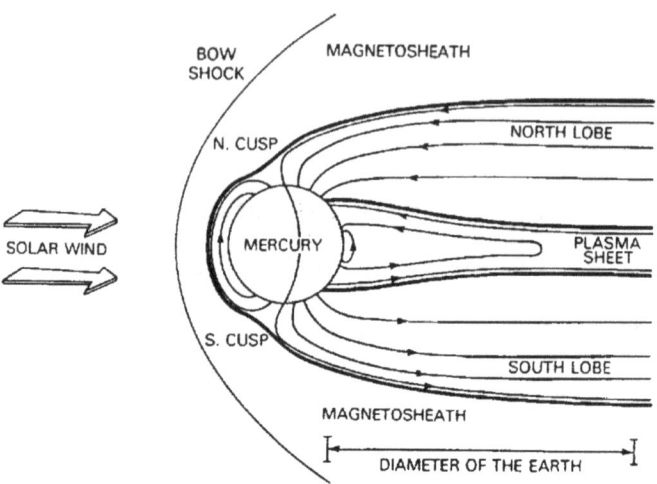

Figure 1. The Mariner-10 flybys of Mercury revealed a magnetosphere much like that of the Earth.

the planet's sunward surface. The two close nightside flybys also revealed a well-developed magnetotail as sketched in Figure 1. Overall, the in-situ observations suggest that Mercury has a magnetosphere whose gross configuration is similar to that of Earth. However, it differs from Earth's magnetosphere in several important aspects that make its further study critical to our understanding of magnetospheric processes, including those responsible for acquisition and dissipation of solar wind plasma and energy.

Mercury's magnetosphere is unique in the following aspects:

A) Mercury has a very tenuous atmosphere and lacks an Earth-like ionosphere. Furthermore, its surface is thought to be poorly conducting. While Mercury does have an exosphere which may play a role in some magnetospheric processes, the absence of an Earth-like ionosphere should allow for the testing of theories regarding the role of ionospheric conductivity in both steady (quiet-time convection) and sporadic (substorm) magnetospheric dynamics. In particular, the coupled interaction between a variable-resistance ionosphere and the nightside magnetosphere is required to initiate substorms according to many theories. If such theories were correct, then substorms would not be expected at Mercury. Other models predict that the resistive nature of the Hermean crust will result in a much larger relative influx of energy from the solar wind than observed at the Earth with brief, intense substorms occurring every few minutes.

B) The linear dimensions of Mercury's magnetosphere are only about 1/7 of those of Earth's magnetosphere. Finite gyroradius effects may, therefore, play a far greater role at Mercury than in any other magnetosphere. Signal transit times and scale sizes for magnetospheric regions and structures are expected to be small, allowing comprehensive spacecraft surveys of magnetospheric phenomena to be carried out far more rapidly than at Earth.

C) Mercury orbits the Sun at distances of only 0.3 to 0.5 AU. As a result of its proximity to the Sun, this planet experiences a significantly different solar wind and interplanetary magnetic field (IMF) than does the Earth's magnetosphere. Observing the response of Mercury's magnetosphere to these extreme driving conditions will allow for the testing of existing theoretical models of plasma interactions and the development of new ones.

In the following sections, we elaborate on ways in which the unique aspects of Mercury's magnetosphere can be exploited by a Mercury orbiter to address fundamental issues in magnetosphere physics.

Absence of an Ionosphere

Mercury has a tenuous, neutral atmosphere whose constituents are poorly known. It is more properly termed an exosphere because the atmosphere is collisionless and the exobase is at the surface; i.e., an atmospheric neutral will typically fall back to the surface of Mercury before colliding with another neutral. The five known species in Mercury's exosphere--H, He, O, Na, and K--are also thought to be important constituents of lunar atmosphere. Mariner-10 ultraviolet spectrometer observations detected H, He, and O at Mercury, while Na and K were later discovered by ground-based optical spectrophotometry. The mechanisms responsible for maintaining an atmosphere at Mercury, despite its high dayside surface temperature and low surface gravity, are not well understood. Atmospheric neutrals must be continually falling onto the surface and being re-emitted from it. Surface interactions are, therefore, critical in determining the atmospheric temperature, composition, and geographic distribution. Magnetospheric processes, including ion precipitation onto Mercury's surface and the pickup of photo-ions, may be extremely important for both atmospheric sources and losses. The term "pickup" is used to describe the process whereby newly created ions are immediately swept away under the convective action of magnetospheric electric and magnetic fields.

The absence of a collisional ionosphere has important consequences for global electric currents and plasma circulation patterns at Mercury. At Earth, high-latitude magnetospheric current systems close by flowing through the ionosphere. At Mercury, these currents cannot close through a collisional ionosphere since none is present, or through the surface because it is expected to be a nearly perfect insulator. One measure of the ability to conduct electricity is the height-integrated Pedersen conductance, which is 1-10 mho at the Earth. The electrical conductivity in Earth's ionosphere is limited mainly by ion-neutral collisions which are essentially absent at Mercury. The ionized component of Mercury's exosphere might be weakly conducting because of the pickup process following photoionization of the different neutral species. Each time a sodium atom, for example, is ionized and picked up, it contributes to a current perpendicular to the magnetic field along the electric field direction. This occurs because each new ion is displaced along the electric field by one gyroradius. The Pedersen conductance at Mercury associated with this mechanism could be as great as ~0.1 mho, but this is still one to two orders of magnitude less than for the Earth's ionosphere.

Closure of magnetospheric current systems through a resistive regolith or partially through an ionized exosphere rather than a collisional ionosphere would have important implications, both for the global current systems and magnetospheric convection as well as for dynamical processes such as substorms and flux transfer events. For example, some theories hold that the timescale for the substorm growth phase at Earth is determined by ionospheric line-tying which limits the rate of magnetic flux return to the dayside following enhanced reconnection at the dayside magnetopause. Other examples are the theories of substorm expansion phase at Earth that consider active feedback between the magnetosphere and ionosphere (specifically enhanced conductivities in the auroral zones) as the essential ingredient for substorms with reconnection and plasmoid formation in the plasma sheet as consequences rather than causes. Such feedback is presumably absent at Mercury. Observations in Mercury's magnetosphere may provide a critical test of these and other models of Earth's magnetosphere.

The Smallness of Mercury's Magnetosphere

In comparing Mercury's magnetosphere with that of Earth, lengths should scale as the planetocentric distance to the subsolar magnetopause. Thus, in units of planetary radii (i.e., R_M or R_E) the implied scaling is $L_M/L_E \sim$ 1.5/10 or $\sim$ 1/7. The first flyby of Mariner-10, past Mercury's nightside, revealed magnetospheric dimensions that generally supported such a scaling. For example, the magnetotail diameter was found to be about $\sim$ 5 R_M as compared with 35-40 R_E at Earth. This linear scaling also was found to apply to other features such as the planetocentric distance to the inner edge of the cross-tail current sheet. This nightside current layer stands off from the Earth by a distance approximately equal to the distance of the dayside magnetopause, and moves 10-20% closer during substorms. The prediction at Mercury would be that this current layer should penetrate within 0.5 R_M of the planet's surface and this was indeed observed by Mariner-10.

The miniature dimensions of Mercury's magnetosphere offer opportunities to study magnetospheric scale length and response-time regimes that will probably never be accessible for observation at any other magnetosphere in the Solar System. Figure 2a shows the near-planet regions of Earth's and Mercury's magnetospheres. Most satellite exploration of Earth's magnetosphere has been done earthward of the Moon's orbit; i.e., within 1.5 tail-diameters (TD) of Earth. Missions such as IMP-8 and ISEE-3 have extended the envelope of exploration to 6 TD. It is known from the early Pioneer-6, 7 and 8 missions that the Earth's tail extends to 100 TD, as shown in Figure 2b, but the sketchy measurements returned were not sufficient to characterize the properties of the magnetotail at these distances. Exploration of the far tail at Mercury will be far easier to conduct by virtue of the relatively distant location of the L2 Lagrange point near 14 TD (cf., 6 TD at Earth).

It is essential to learn how magnetospheric structure evolves at large distances and how the magnetotail responds to changes in the interplanetary medium. Does the magnetotail have a coherent structure; i.e., an identifiable plasma sheet and lobes, which extends to very large distances (> 10-20 TD)? Does the magnetotail "flap" like a flag at large distances due to the Kelvin-Helmholtz instability? If a mission to Mercury employed two spacecraft, then it might be possible, for example, to capture one of them in the L2 orbit while the other was put in Mercury orbit as illustrated in Figure 2b, or execute other distant tail trajectories similar to those utilized by ISEE-3 at Earth. Structure and motion of the very distant tail could then be related to solar wind and near-planet magnetosphere changes. The short solar wind travel time (7 minutes) from Mercury to the L2 orbit results in a close coupling, making cause-and-effect relations easier to discern than at Earth. Many workers believe

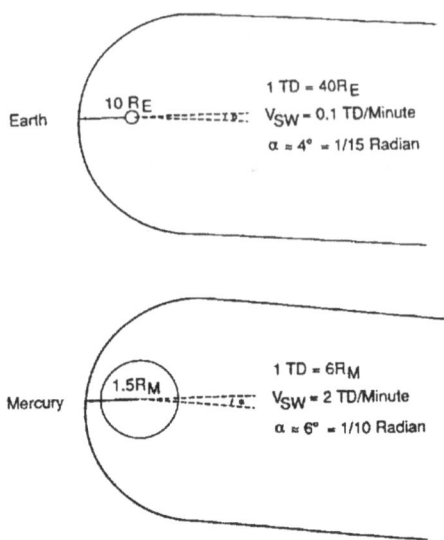

Figure 2a. Equatorial cross sections of the near-planet regions of Earth's and Mercury's magnetospheres. The planets are shown scaled correctly relative to their magnetospheres. For Earth the tail diameter (TD) is 40 $R_E \sim$ 255000 km, while for Mercury 1 TD $\sim 6R_M \sim$ 14700 km. The average velocity of the solar wind (V_{SW}) is given in units of TD/minute. The aberration angle, α, is the mean angle of solar wind flow, relative to the planet-Sun direction, due to the planet's motion around the Sun.

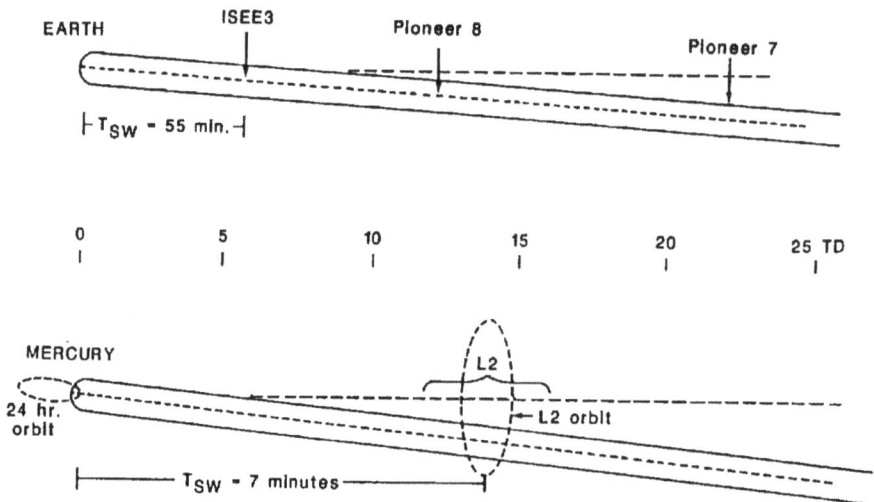

Figure 2b. Extensions of Earth's and Mercury's magnetospheres to 25 TD are displayed. For Mercury a 24-hour elliptical orbit is depicted as well as a possible orbit around the L2 point. A satellite in the L2 orbit would remain anti-sunward of Mercury and near or within the magnetotail as Mercury moved around the Sun. Solar wind travel times (T_{SW}) along portions of the magnetotails are shown.

that during substorms in Earth's magnetosphere, the plasma sheet is severed by magnetic reconnection quite close to Earth and flows rapidly down the tail as a magnetically confined structure called a plasmoid. Some theories predict that this is the primary way that solar wind plasma and energy, earlier acquired by the magnetosphere, is dissipated and a portion returned to the solar wind. Thus, plasmoids may be of fundamental importance to magnetospheric physics. Predictions of the occurrence of plasmoids were first made based on magnetotail observations within ~ 1 TD of Earth, and strong support has come from ISEE measurements at ~ 6 TD. Observations of plasmoids in Mercury's distant magnetotail would provide important confirmation that magnetic reconnection and plasmoid formation are basic features of the process by which stored energy is released within planetary magnetospheres.

Solar wind energy coupling into Earth's magnetosphere is known to be strongly influenced by the polarity of the IMF. Southward IMF leads to strong coupling, through reconnection with the northward geomagnetic field at the surface of magnetosphere. The occurrence of substorms, the basic mechanism for stored energy release and dissipation, clearly relates on a statistical basis to the occurrence of southward IMF. However, IMF direction typically varies on a time scale of a few minutes; much shorter than the time scale of energy storage and substorm occurrence at the Earth which is about an hour. Accordingly, detailed cause and effect relationships are very difficult to discern. In the case of Mercury, where the magnetospheric response time is believed, on the basis of Mariner-10 data, to be only a minute or so, relations between the IMF and internal magnetospheric processes could be studied with great benefit. For example, it is not uncommon for the IMF to remain southward and constant for ten minutes. At Mercury this time span is long compared to the substorm cycle time and it would be possible to see whether the magnetosphere responded to this situation by repeated substorms and plasmoid releases as some substorm theories predict. In order to conduct such an experiment at Earth the IMF would have to remain southward for 1 to 2 days, a highly improbable occurrence. Furthermore, even if the IMF requirements were satisfied, the orbital periods for magnetospheric trajectories at Earth, hours to days, would carry the observing spacecraft in and out of the key regions over the course of event and complicate or prevent the determination of causal relationships.

Summary

Mercury is the best place to test and extend the understanding of magnetospheric physics acquired by studying the Earth's magnetosphere. The major difference between Mercury and Earth, the former's lack of an ionosphere, is highly valuable in that it will allow the testing of terrestrial theories in the limit of large (or infinite) polar cap electrical resistivity. Mercury's small magnetosphere may also solve the space-time ambiguity problem that has confounded efforts to perform synoptic studies of Earth's magnetosphere. Approximately once per hour the solar wind conditions change significantly, and magnetospheres must change to accommodate these new conditions. An Earth satellite takes many hours to a day to traverse each of the magnetosphere's structural units, which in the meantime is changing its shape and behavior. At Earth a satellite virtually never samples a complete structural unit before it changes its state. Hence, a statistical approach is necessary for synoptic studies of

Earth's magnetosphere. A satellite at Mercury crosses the entire magnetosphere in one-third of an hour or less. Most often the solar wind will not change during this time. Thus, the changes a satellite records in a magnetospheric structure at Mercury characterizes that structure while the magnetosphere is in a fixed state. Mercury's small magnetosphere may also contribute to the elimination of the confusion between solar wind and magnetospheric convection time scales that exists for the Earth's magnetosphere. Simple scaling arguments put the convection time scale of Mercury's magnetosphere around a few minutes. The convection time scale of Earth's magnetosphere is around one hour, essentially the same as the characteristic time for significant solar wind changes.

In summary, there are three primary reasons for going to Mercury to extend our knowledge of magnetospheric physics. 1) Of all known magnetospheres, the investigation of Mercury's is the most likely to produce new knowledge which is directly applicable to Earth's. 2) The lack of a conducting ionosphere and the solar wind conditions at 0.3-0.5 AU place boundary conditions on Mercury's magnetosphere which are significantly different from those influencing the Earth's magnetosphere. 3) Mercury's magnetosphere is of a dramatically smaller scale size than the terrestrial magnetosphere; a factor of major importance in separating temporal and spatial variations.

2.2 PLANETOLOGY

Our current knowledge of Mercury is based almost exclusively on the data received from the Mariner-10 spacecraft on its three flybys of Mercury in 1974-75. Mercury represents an end member in Solar System origin and evolution in that it formed closer to the Sun than any other planet and, therefore, in the hottest part of the solar nebula. Until we have an understanding of the properties, composition, and history of Mercury, we will not have a complete understanding of Solar System formation.

Mercury has several unique properties including its high mean density of 5.44 g/cm^3. This indicates that it has a larger weight percentage of iron than any other planet or satellite in the Solar System and thus its iron core radius is some 75 percent of the planet radius, or about 42 percent of the planet volume. The presence of a dipole magnetic field indicates that at least a fraction of this large core is still in a fluid state. Mercury also has the largest orbital eccentricity (.205) and inclination (7 degrees) of any planet other than Pluto. Its rotation period (58.6 days) and orbital period (88 days) are in a 3:2 resonance so that Mercury makes three rotations around its axis every two orbits around the Sun. The slow rotation period and close proximity to the Sun (.387 AU) result in the largest range of surface temperatures in the Solar System (-183° to +427°). Recent Earth-based observations have shown that Mercury is surrounded by a tenuous atmosphere that is probably derived from its surface.

Mariner-10 imaged only about 45 percent of the surface at an average resolution of about 1 km, and less than 1 percent at resolutions between about 100 m to 500 m. Furthermore, about half of this coverage was at high Sun angles, rendering topographic discrimination difficult or impossible. This coverage and resolution is somewhat comparable to Earth-based telescopic coverage and resolution of the Moon before the advent of spaceflight. Our knowledge of Mercury, therefore, is at about the same level as our knowledge of the Moon in the early 1960s.

Mercury's surface superficially resembles that of the Moon. Like the Moon, it displays ancient, heavily cratered highlands and younger, smooth plains largely concentrated within and surrounding large impact basins. Mercury's surface, however, has characteristics that set it apart from the Moon and other terrestrial planets. A presumably global distribution of lobate scarps (thrust faults) attests to a unique tectonic framework that was probably caused by crustal shortening resulting from planetary contraction due to cooling of the core and lithosphere. Unlike the Moon, the major terrain type is old intercrater plains interspersed between craters in the highlands. Furthermore, the albedo of surface units on Mercury is significantly higher than comparable units on the Moon, suggesting a different composition. The plains units (intercrater and smooth) have been interpreted to be volcanic, but this interpretation is uncertain due to a lack of good photographic coverage and resolution, and compositional information. In spite of the large amount of new knowledge gained from Mariner-10, we still know very little about Mercury's surface, interior, composition, and geologic history.

Important Questions About Mercury

There are a number of important questions about Mercury which must be addressed before we fully understand the planet's formational and geological history. To address fully these questions requires an orbiter with a complement of instruments and an orbital configuration optimized for planetological observations. Ideally this would consist of a three-axis stabilized spacecraft in near-circular polar orbit, but at present this approach is not feasible. Spinning spacecraft, however, can operate in Mercury's harsh thermal environment and return important new information with spin-scan imaging and surface composition instruments. In addition, Doppler radio tracking can provide information on the local gravity field. The mission should allow global coverage at a resolution of at least 1 km/pixel and about 25 percent coverage at 500m/pixel or better. A minimal requirement of the instruments which measure surface composition (X- and Gamma-Ray Spectrometers) is the accurate determination of global abundances of Si, Mg, Fe, Al, Ti, K, Th, Na, and Ca. If these data can be obtained, then important new insights can be obtained on the problems discussed below.

Origin: The origin of Mercury and how it acquired such a large fraction of iron compared to the other terrestrial planets is not understood. Equilibrium condensation models for Mercury's position in that part of the solar nebula cannot account for the large fraction of iron which must be present to explain its high density. These models suggest that the maximum amount of iron that could be condensed and acquired from other feeding zones is only enough to result in a mean uncompressed density of about 4.2 g/cm^3, rather than the observed 5.3 g/cm^3. Until this problem is resolved, we will not fully understand how the terrestrial planets formed.

Three hypotheses have been put forward to explain this discrepancy. One (selective accretion) involves an enrichment of iron due to mechanical and dynamical accretion processes, while the other two (post-accretion vaporization and giant impact) invoke removal of a large fraction of the silicate mantle from a once larger proto-Mercury. All three hypotheses have their strengths and weaknesses, but currently there are few data to test them. Fortunately, each hypothesis predicts a significantly different composition for Mercury's silicate fraction, which could be inferred from compositional information obtained by a geochemical experiment package flown on a Mercury orbiter.

In the selective accretion model, the differential response of iron and silicates to impact fragmentation and aerodynamic sorting leads to iron enrichment due to the higher gas density and shorter dynamical time scales in the innermost part of the solar nebula. As a consequence, the silicate fraction should have a refractory oxide abundance of between about 7 and 9 weight percent (alumina ~3.6 to 4.5 percent, relative enrichment), alkali oxides about 1 weight percent (relative enrichment Na, K), and FeO between 0.5 and 6 weight percent (relative enrichment Fe).

In the "Giant Impact" hypothesis, a planet-sized object impacts Mercury and essentially blasts away much of Mercury's silicate mantle leaving a large iron core. In this case, Mercury is metal-rich because it is small enough that much of the ejected silicates do not reaccrete and are swept up by Venus and Earth. The chemical consequences for Mercury's remaining silicate fraction are about 0.1 to 1 weight-percent refractory oxides (low Al), between 0.01 and 0.1 percent alkali oxides (low Na,K), and 0.5 to 6 percent FeO (enrichment). Vaporization predicts decreases in Fe, enrichment refractories (Al), and low alkalis (Na,K).

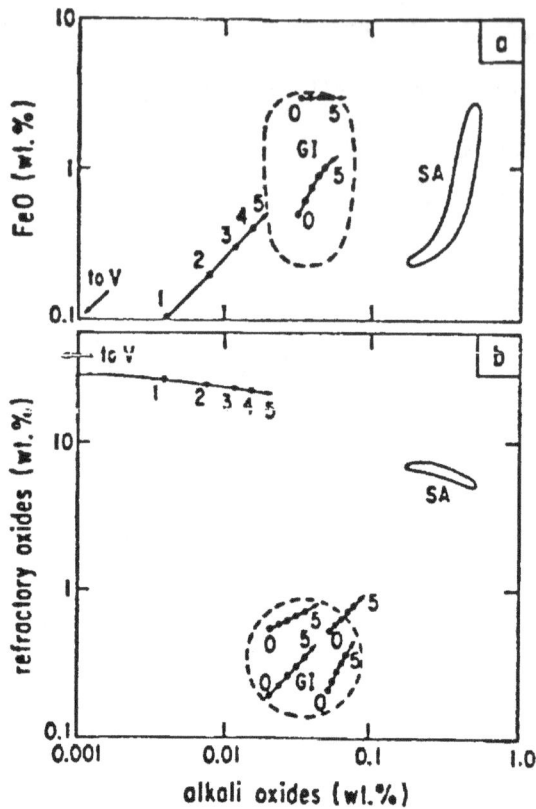

Figure 3. Predicted composition of Mercury

Figure 3 presents a diagram showing the predicted composition of Mercury for the three hypotheses discussed above. Compositional information from a Mercury orbiter could help discriminate between these models, or, at the very least, place severe constraints on any hypothesis invoked to explain Mercury's very large fraction of iron.

Composition: Virtually nothing is known about the composition of Mercury other than that it must have a large fraction of iron that is probably concentrated in an enormous iron core. The tenuous atmosphere surrounding Mercury indicates that these elements are present on the surface of Mercury, but their abundance there is unknown. Earth-based spectroscopic data for Mercury's surface composition are poor, owing to the very poor observing conditions, and are contradictory. They do suggest, however, that compositional variations occur across the surface. Color differences between the Moon and Mercury have been inferred to mean that the surface of Mercury is depleted in iron and titanium relative to the Moon. However, we have no knowledge of the major- or trace-element abundances of the Hermean surface and how they differ from those of the Moon. Are there compositional variations in space and time across the surface that correlate with geologic units? (Color-difference data suggest a lack of correlation.) What are the implications of surface composition for crustal, mantle, and core compositions? Is there evidence for the existence of a global magma ocean? If so, are the differentiation products different from those on the Moon? Global geochemical maps, coupled with gravity data, and imaging should help to generate answers to these questions, such as providing evidence for the existence, depth, extent, and differentiation products of a global magma ocean, compositional variations in space and time of igneous processes, and estimates of crustal thickness and density variations.

Magmatic History: One of the outstanding problems of Hermean geology is the origin and extent of plains units. Are they entirely volcanic or are they some mixture of lavas and impact ejecta? What was the nature and origin of highland and lowland igneous activity through time? What have been the mechanisms of heat transfer in the interior over geologic time? How did the lithosphere thickness vary spatially and through time, and what were the accompanying tectonic styles at the surface? Global geochemical maps and imaging should permit the identification of rock types and the origin(s) of plains units. These data should provide insight into the nature, compositional variations, and duration of highland and lowland igneous activity and the discrimination between volcanic and impact products.

Crustal Dynamics (Tectonics), Thermal History and Internal Structure: The tectonic framework of Mercury is unique in the Solar System and appears to be dominated by thrust faults that may have a global distribution. Because of the limited photographic coverage, however, we do not know how widespread the distribution of these structures is, and whether there are major regions of extensional tectonics. A major goal of a Mercury orbiter will be to determine the global distribution of this unique tectonic framework and its development through time. If this system of thrust faults is global and is the result of crustal shortening due to cooling of the lithosphere and core as the current evidence indicates, then a unique opportunity exists to set tight constraints on the present thickness of the fluid core.

A currently molten outer core requires a light alloying element in the core to lower the melting point and retain a partially molten core throughout geologic history. Although oxygen is such an element, it is not sufficiently soluble in iron at Mercury's low internal pressures and, therefore, sulfur is the most reasonable candidate. The present extent of the outer molten core and the onset of solid inner core formation are highly dependent on the abundance of sulfur in the core. For a sulfur abundance less than 0.2 percent, the entire core should be solidified at the present time, while an abundance of 7 percent results in an entirely fluid core at present. Inner core formation begins about 3.9 GY ago for 0.2 percent sulfur and results in an outer fluid core about 100 km thick at present. For 5 percent sulfur, the inner core begins to form about 2 GY ago and results in an outer fluid core about 1150 km thick at present.

Initial chemical equilibrium condensation models for Mercury's position in the solar nebula predicted the complete absence of sulfur, which is inconsistent with the presence of a partially molten core suggested by the dipole magnetic field. However, three-dimensional (3-D) simulations show that considerable mixing from the feeding zones of other terrestrial planets can occur during the accumulation of the terrestrial planets, and thus alter their initial compositions. As a consequence, equilibrium condensation models for Mercury have been modified so that 60-90 percent of the material is accreted at Mercury's present distance while 10-40 percent comes from planetesimals perturbed from the feeding zones of other terrestrial planets. This could supply from 0.1 to 3 percent FeS depending on the degree of mixing.

Accurate measurements of scarp lengths and heights from a Mercury orbiter can be used to determine the amount of crustal shortening due to cooling of the lithosphere and core. This, together with equations-of-state for Mercury's interior, can be used to set limits on the thickness of the fluid outer core and, therefore, the sulfur content. Limits on the sulfur content can then constrain the degree of planetesimal mixing involved in Mercury's formation. Present estimates based on estimates of the scarp heights and lengths over about 25 percent of the surface and extrapolated to the entire planet suggest a fluid outer core thickness of about 900 km, but this is highly uncertain because of the poor surface coverage and resolution from the Mariner flybys.

Another problem that can be addressed by a study of lobate scarps is the onset of global contraction and the amount of radius decrease. Current thermal history models predict that contraction began immediately following accretion and has resulted in a radius decrease of about 6 to 8 km. However, the thrust faults observed in the areas imaged by Mariner-10 appear to post-date intercrater plains formation suggesting that the onset of planetary contraction began relatively late in mercurian history, and that the amount of global contraction was only about 2 km.

These estimates of global contraction and fluid core thickness must remain very tentative because of the limited coverage and resolution of Mariner-10 imagery (one hemisphere). Detailed mapping of the scarps and their transection relationships from Mercury Orbiter data can place tighter constraints on the onset and amount of global contraction, and therefore, better define thermal history models.

Finally, imaging of tectonic features associated with lithospheric loading (e.g., in the Caloris Basin) and gravity data should permit investigation of variations in lithospheric thickness. These observations, together with observations of igneous processes through time, may be used as further constraints on the thermal history of Mercury.

Impact Processes: Imaging, compositional information, and gravity data will enable studies of crater and basin structure, morphology and composition of ejecta deposits to reconstruct pre-impact target composition and structure, and post-impact ejecta deposition and modification. Imaging will greatly improve impact crater statistics and their spatial variations to reconstruct accurately Mercury's geologic and cratering history.

Gravity Field: The local gravity field derived from Doppler tracking and combined with imaging and topographic data will determine whether gravity anomalies are associated with topography. If so, these data can be used to derive lithospheric thickness and to determine whether mascons exist and are associated with impact basin fill as they are on the Moon.

Magnetic Fields: The discovery of a global magnetic field during the Mariner-10 encounter with Mercury was a surprise. Although the equatorial surface field is only ~300 nT, or about 1% of the Earth's field, it is generally considered too large not to be caused by a presently active internal dynamo. It is almost certainly the weakest dynamo in the Solar System and, as such, represents a unique challenge to theory. The Hermean dynamo is weak presumably because of the slow rate of rotation and relatively small size of the planet. However, Mercury also has a unique internal structure consisting of a large metallic core surrounded by a relatively thin liquid shell. Other unique features of the planet may also play a significant role.

The principal obstacle to fitting Mercury into a general scheme of planetary magnetic fields is the extremely limited information that is available. The two passes of Mariner-10 by Mercury were subject to severe limitations. The magnetosphere is so small that only a restricted range of radial distances was potentially available for study. The planet also rotates slowly, with a period of ~59 days, so that the longitude range covered by the observations was also very limited. Finally, the scale of the magnetosphere, and its high level of magnetic activity, meant that the field measurements included a large contribution from nearby, time-varying magnetospheric currents which are basically unrelated to the dynamo currents inside the planet.

These limitations have made it impossible to carry out the usual spherical harmonic expansion of the planetary field with the degree of confidence achieved for Jupiter and the outer planets. Because of the trajectory limitations, the ratios of the dipole, quadrupole and octupole moments inside Mercury may be adjusted arbitrarily to achieve a fit of the models to the data. In describing planetary fields the low-order dipole moment usually receives most of the attention. However, the absence of accurate knowledge regarding the higher order moments is a serious hindrance in studying the field at the surface of the core and in assessing the relative contribution of the higher order field components. The latter information is related to the spectrum of convective motions in the source region.

Another aspect of planetary field investigations which needs attention at Mercury is the possible presence of crustal magnetic anomalies. Such anomalies have been detected at the Earth's surface and as magcons on the surface of the Moon where they are correlated with impact craters. Do such features occur on the lunar-like surface of Mercury? Crustal anomalies are also of interest because of their possible correlation with heat flux anomalies. Correlations with gravitational and thermal anomalies can provide significant information about the regolith, such as variations in composition and thickness.

Summary

A Mercury orbiter mission has the potential of providing important new information about Mercury assuming it carries an imager, instruments to measure the surface composition, and Doppler radio tracking. Global imaging at a resolution of about 1 km/pixel, and about 25 percent coverage at 100 m/pixel resolution or better is required. Furthermore, much of the surface should be viewed under a variety of phase angles, providing valuable photometric information; fine-scale surface structure, albedos, and photoclinometry. In addition, stereo coverage from images taken at different viewing angles would be extremely valuable. This could provide crucial quantitative topographic information from which topographic maps can be prepared. The surface elemental abundances of Si, Mg, Fe, Al, Ti, K, Na, Th, and Ca can be determined with a gamma-ray spectrometer and x-ray fluorescence experiment, but the accuracy and surface resolution depend on the orbiter spin rate. In addition, Doppler radio tracking should allow the determination of the local gravity field. If these data are obtained, then important new insights can be obtained about the origin of Mercury and its composition, crustal dynamics, internal constitution, magmatic processes and history, impact processes, and geologic and geophysical history.

III. HELIOSPHERIC AND SOLAR SCIENCE OBJECTIVES

3.1 PHYSICS OF THE INNER HELIOSPHERE

A Mercury orbiter mission would provide a unique and extremely valuable opportunity to study the plasma physics of the inner heliosphere. As the logical follow-on to Helios-1/2, MeO would allow the existing data base to be extended with more comprehensive measurements possessing significantly better spatial and temporal resolution. If dual orbiters were flown, then simultaneous observations over a short baseline could be obtained. This would result in major gains in scientific knowledge comparable to that achieved in magnetospheric physics by the highly successful, multi-spacecraft ISEE mission.

The solar wind investigations on Mercury orbiter should be oriented towards the following scientific goals: (1) to identify the plasma processes that are central to the origin of the solar wind; (2) to identify the processes that control the evolution and dynamics of the solar wind; and (3) to determine the solar wind parameters that influence the dynamics of Mercury's magnetosphere.

These goals should be achieved by measuring the distribution functions of electrons and major ion species together with the interplanetary magnetic field and plasma waves. The required set of state-of-the-art particle sensors, as combined in the solar wind plasma analyzer, must render possible an unambiguous separation of the major plasma constituents, a complete coverage of their distributions in 3-D velocity space, high time resolution, and a precise tracing of the electron "strahl" as a probe of the outer coronal and interplanetary magnetic field topology. Solar wind in-situ measurements from 0.3 AU (Helios) out to beyond 50 AU (Pioneer-10/11) have already yielded a rich harvest of information on the state of the heliospheric plasma and local processes, and on the dynamical evolution of the solar wind and structures embedded therein. However, many questions remain unanswered and basic problems, unsolved. Among those are the acceleration of high-speed flows emanating from coronal holes, the identification of the source regions for the low-speed wind and its acceleration mechanism, the structure of heliospheric current sheet flows and magnetic "clouds" found therein, the generation and effects of MHD turbulence in various coronal sources, and the interplanetary dynamical evolution of the resulting turbulence thereafter and its possible dissipation. The structure and dynamics of small-scale current layers, like shocks and discontinuities, are still matters of intensive, ongoing research. Transport problems in the weakly collisional and micro-turbulent solar wind plasma regime are entirely unresolved--a statement which also applies to coronal plasma physics.

Coronal Sources and Large-Scale Structures of the Solar Wind

On the largest scales, the various categories of solar wind plasma flow and magnetic field structures are related to or strongly determined by the large-scale physical conditions in the solar corona. The close relationship between the coronal magnetic field structures and plasma flow properties was revealed during the celebrated Skylab mission. In particular, the Skylab images showed the distinct role of corotating coronal holes as the sources of recurrent high-speed streams, and the close association of coronal streamers with the heliospheric current sheet. Many observations already exist regarding the longest term and largest scale variations in the structure of the solar wind, and much effort has been spent to seek consistency between the interplanetary observations and those of the corona. It has become increasingly clear, however, that all in-situ observations reflect only special conditions and particular phases of the solar cycle. The observations need to be interpreted in terms of the natural variations occurring in the inner heliosphere as a result of solar activity and the periodic restructuring of the coronal magnetic field.

In contrast to high speed streams, the coronal sources of slow solar wind still remain obscure. Mapping back of streamlines to the Sun leads us to identify their sources roughly with the closed magnetic structures of the streamers associated with the magnetic neutral line. There must be a transition from closed coronal field lines to open interplanetary field lines while they are dragged away by the outflowing plasma. Consequently, the loops themselves are drawn out and should open somewhere, or else be cut off from their foot points, a process implying magnetic reconnection if a continuous build-up of solar magnetic flux is to be avoided. Some of the magnetic "clouds" or plasmoids observed in interplanetary space might have originated from such a process.

The slow-speed solar wind also appears to be much more inhomogeneous and of more intermittent nature than long-lasting fast streams. This suggests that the magnetic topology of its source region plays a crucial role in determining the flow characteristics. For example, the helium content of the slow wind is variable and lower than in fast streams. This depletion might indicate that the plasma emanated from higher, spatially variable layers in the solar atmosphere. Extreme plasma conditions are usually observed in the heliospheric current sheet which is embedded in slow solar wind. There the electrons, as well as ions, tend to be comparatively cold and they cool off almost adiabatically with heliocentric distance. As a result, Coulomb collisions were found to be important and even able to effectively limit ion temperature and velocity differences. The current sheet and its neighboring plasma proved to be the proper collisional domain of the solar wind, while high-speed streams are collisionless. To identify the coronal sources of the current sheet plasma and associated plasma conditions which give rise to these distinct interplanetary characteristics, is an important issue to be addressed by the Mercury orbiter.

MHD Turbulence

At intermediate time scales of a few hours and below, the solar wind plasma appears to be in a highly turbulent state composed of many different components. The predominant component is Alfvénic fluctuations, which under certain conditions are exact nonlinear solutions of the ideal MHD equations and, in spite of their large amplitudes, resemble linear Alfvén waves. The second component of the turbulence has a lower energy density than the Alfvénic fluctuations and is compressible. It appears to be enhanced in mixed low-speed flows and at travelling interplanetary shocks. Several important plasma-physical questions are related to the observed turbulence.

The solar wind has been described as a turbulent, incompressible magnetofluid, a hypothesis which has been tested by measuring the so-called rugged invariants; i.e., the total energy and magnetic helicity of the fluctuations. This approach has led to significant advances. However, many basic questions remain open, for example the possible relation between the compressible and incompressible parts of the turbulence. Large-amplitude Alfvén waves are prone to parametric instabilities limiting their amplitudes. Alfvén waves may steepen to form rotational discontinuities in the presence of density fluctuations, which in turn, represent a source or sink term in the rate-of-change equation for the cross helicity. This is then a variable and will change whenever the flow and Alfvén velocity are divergent. As a consequence, outwardly and inwardly propagating Alfvén waves will interact with each other and wave energy will cascade and spread out in Fourier space.

Recent numerical simulations indicate that power spectral indices for waves of opposite propagation directions may be different. A related fundamental question is whether the Alfvénic fluctuations are produced locally, with a sufficiently nonlinear amplitude to interact significantly among themselves, or whether the waves are merely produced in the lower solar corona and propagate freely outward. Observed wave amplitudes are such that the ratio of interaction time over eddy turn-over time is rather large. It is, therefore, necessary to question the scenario in which the interplanetary turbulence at ~1 AU is considered to be fully developed. Observations from a Mercury Orbiter at 0.3 to 0.5 AU will add greatly to the resolution of this issue.

Plasma Kinetics

After more than two decades of in-situ plasma observations of ion and electron velocity distributions, a rather complex picture of the interplanetary plasma state has emerged. The spatial inhomogeneity of the solar wind, the variability of the solar boundaries and plasma sources, and the complex topology of the coronal and interplanetary magnetic field, all manifest themselves in a variety of nonthermal features of the velocity distributions. Small-scale dissipative processes, Coulomb collisions, and kinetic wave-particle interactions all shape the distributions. Electron and ion velocity distributions, as a rule, strongly deviate from Maxwellians because particles react to the average, large-scale forces and also to the small-scale forces associated with waves.

Helios observations have provided a fairly complete phenomenological picture of electron distributions in various solar wind flows at distances between 0.3 and 1 AU. On a broad statistical basis the usual core-halo structure has been confirmed by discerning a persistent break in the energy spectra; i.e., a sudden change in the slope of the distribution. This varying break point energy may relate to the interplanetary electrostatic potential or be determined by Coulomb collisions.

In recent years the need for an understanding of the role of Coulomb collisions in the solar wind has steadily increased. Primarily, it is necessary to better understand the electron transport phenomena in the corona and the solar wind, in the context of thermally driven solar wind flows. Since classical transport theory does not apply, new kinetic concepts, replacing local laws by local and global relations, have been developed and are under study. Some studies have employed the full Fokker-Planck collision operator to kinetically model electron heat conduction in the solar transition zone.

Similarly, the important role Coulomb collisions play in the evolution of ion velocity distributions has been established in various papers. It has been shown in numerical simulations that the combined action of the mirror effect (diverging IMF) and collisional scattering (run-away) can produce skewed, and even double-peaked, ion distributions, as found in observations. The importance of collisions is therefore without doubt, particularly in flows at and close to the heliospheric current sheet, where collisional transfer of energy and momentum between alpha particles and protons is rather effective.

To firmly establish the role of collisions in space plasmas, where thermal free paths of particles are usually comparable with fluid expansion lengths or system dimensions, knowledge of the full three-dimensional velocity distribution is indispensable. The advanced capabilities of solar wind instruments conceived since the Helios-era promise further progress in the area of nonclassical collisional transport when MeO observations in the inner Solar System become available.

Plasma Waves

The plasma wave experiment should permit measurements in the frequency range between 10^{-1} and 10^5 Hz of magneto-acoustic waves, ion-cyclotron waves, ion-acoustic and whistler mode turbulence, and electron plasma oscillations. The observations of these different wave modes are required to analyze the overall state of turbulence of the interplanetary medium, the stability of electron and ion distribution functions, the stability of solar wind microstructures such as current sheets and shock waves, the propagation of energetic electrons, and the onset and temporal evolution of plasma kinetic instabilities. These waves play a fundamental role in the overall internal energy budget and transport of the solar wind; e.g., in the ion-electron drag and the coupling between different ion species, and therefore largely determine the fluid-like behavior of the macroscopic solar wind. Measurements of these waves in connection with detailed plasma observations are therefore to be carried out throughout the solar wind cruise phase of the Mercury orbiter mission as described below.

Solar wind proton "double" streams and proton-alpha differential streaming can provide sufficient free energy to drive field-aligned magnetosonic waves. Statistical analysis of Helios data have provided evidence for the marginal instability of these beams with the largest growth rates occurring in high-speed streams. Typical growth times of the instability were estimated to be several tens to hundreds of seconds. In this case, the distribution should relax quasi-linearly by slowing down and diffusing the beams in pitch angle. High-time resolution plasma wave measurements taken by MeO in the inner heliosphere would allow us to follow the details of this relaxation process, and the subsequent evolution of the beam instability through its nonlinear stage. The full understanding of such ion beam instabilities is a primary objective for the space plasma physics community.

Ion-cyclotron and magnetosonic waves have been separately resolved using magnetic polarization measurements performed well within 1 AU. As these waves play an important role in the acceleration of minor ions, proton heating, and double-beam evolution, the study of these waves and of the instabilities driving them is of great importance especially in the inner heliosphere where the waves are expected to be more intense than at 1 AU.

Whistler mode turbulence occurs not only in high-speed solar wind streams (as Alfvén waves predominantly do), but also under normal and quiet solar wind conditions. Studies in this area should focus on investigating the origin of this type of turbulence. Thermal anisotropies in the ion distribution functions and instabilities caused by electron heat fluxes have been proposed as candidates that excite the turbulence. However, these processes have not yet been observed or actually demonstrated to be operating in the interplanetary medium.

Ion acoustic waves are believed to be driven unstable by double-ion streams and by the electron heat flux instability, or by weak currents provided $T_e \gg T_i$. As currents occur in shocks, current sheets and discontinuities, the observation of these waves in connection with microscopic solar wind structures is of great importance. Since most of these microscopic structures occur more frequently closer to the Sun, the proposed mission to Mercury is ideally suited for detailed investigations of ion acoustic noise.

Finally, electron plasma oscillations are observed in connection with type-III radio bursts. Helios observations have clearly shown that these waves are generated and controlled by energetic electron beams. Additional observations of these waves still need to be made closer to the 'source' region of both the metric bursts and the electron beams, where effects such as velocity dispersion, scattering, and defocusing of the electrons play only a minor role. MeO will provide such measurements.

3.2 SOLAR FLARES AND SOLAR ENERGETIC PARTICLES

The study of solar flares from the unique vantage point of Mercury orbit offers exciting possibilities. For as-yet-unknown reasons, the Sun from time to time suddenly produces vast bursts of energetic particles and electromagnetic energy in a process called a solar flare. It seems clear that the process is caused by the conversion of some of the large store of magnetic energy in active regions into high-energy particles, both electrons and ions. Exactly how, why, or even precisely where in the Sun's atmosphere the particles are accelerated is not understood. This is a fundamental problem of astrophysics which has defied solution. Besides the importance of understanding how the flare can concentrate energies from a few MeV to several hundred MeV (and even GeV) in individual particles, the output of a flare provides a highly variable energy input into the heliosphere. It is also possible that the Sun produces energetic nuclei and electrons in apparently nonflaring situations. Because of the proximity of Mercury to the Sun, important new observations of flare electromagnetic emissions, x-rays and gamma-rays can be made by MeO. The flux of 100 MeV neutrons from a given flare will be increased by more than the r^2 factor (~10) if observed at Mercury rather than at the Earth; the flux of MeV neutrons will be increased by a factor of ~1000 because of the increased survival probability against radioactive decay. Of equal importance is the study of the solar energetic charged particles, because the spectra and fluxes of these flare emissions are strongly influenced by the interplanetary magnetic field between the Sun and the observing point. Hence, observations from Mercury will more directly reflect the characteristics of these radiations at the flare source.

Our current knowledge on the characteristics of solar flare particle acceleration comes from: 1) hard x-ray and gamma-ray measurements, 2) observations of solar energetic particles (SEP) in the interplanetary medium, 3) direct observations of flare-produced neutrons at 1 AU, 4) observations of solar neutron decay protons, and 5) ground level solar cosmic ray events. These observations have shown that probably all flares simultaneously accelerate both MeV electrons and 50 MeV ions in time scales of a few seconds. Occasionally, intense events occur which reveal the presence of 100 MeV electrons and GeV ions which interact in the solar atmosphere producing very energetic bremsstrahlung, meson gamma-rays, and several hundred MeV neutrons. There is also evidence from some flares that the highest energy ions (i.e., ~300 MeV) which interact at the Sun are continuously produced for as long as 30 minutes after the impulsive flare. From the observations, one hopes to infer the properties of the solar flare acceleration mechanism(s), but presently our detailed knowledge is very limited.

We can attack these basic flare problems from Mercury by making the following observations: 1) the solar flare neutron energy spectrum below 30 MeV and the 2.223 MeV capture gamma-ray line flux, 2) the directivity of hard x-rays and, 3) the spectra and time structure of the solar energetic charged particles as described below.

Observations of Neutrons and Gamma-Ray Lines

The solar neutron and gamma-ray line observations can be used to derive the low energy particle spectrum since both are produced directly by nuclear interactions of flare-accelerated protons and heavier ions with the ambient solar atmosphere. Such interactions are believed to occur in the photosphere or lower chromosphere. The neutrons result primarily from the disintegration of ^{4}He and heavier nuclei and occasionally from p-p collisions. Therefore, the neutron production rate is strongly dependent on the spectral distribution of the accelerated particles. Previous studies with gamma-rays have shown that the nuclear interactions producing solar neutrons are caused by accelerated particles which remain trapped in the magnetic fields of the flare region and interact as they slow down in the solar atmosphere, rather than by accelerated particles which eventually escape into interplanetary space. If the escaping particles were responsible for neutron production there would have been an enrichment of spallation products, such as D, T, Li, Be, and B, which were not observed in the solar energetic particles in interplanetary space.

Neutrons in the solar atmosphere can interact with the ambient gas, decay or escape from the Sun. What actually occurs depends on the energy and spatial distribution of the neutrons. Neutrons initially moving upward from the production region (i.e., lower chromosphere) have a high probability of escaping from the Sun. Downward-moving neutrons have a small probability of escaping through elastic scattering on hydrogen, but limb flares or flares well away from the disk center could give side-scattered neutrons which could reach the Earth. Elastic scattering on the less abundant helium could also scatter neutrons toward the Earth. Most downward moving neutrons are captured by protons in the photosphere after thermalization, giving the characteristic 2.223 MeV gamma-ray line, but nonradiative capture of neutrons on ^{3}He is just as likely. Simultaneous observations of this line and of neutrons $\geq$10 MeV therefore yield information on the depth and directionality of neutron production. Beaming and directionality of flare electrons and protons can be inferred from stereoscopic observations of bremsstrahlung, neutral pion decay gamma-rays and neutrons using widely separated detectors. The potentially large parallax provided by MeO and near-Earth detectors would be invaluable for such observations.

The flux of high energy neutrons at the Earth results directly from the production during the slowing-down phase of those ions that remain trapped at the Sun. Since neutrons are unaffected by solar or interplanetary magnetic fields they can travel directly to the Earth, whereas the ions travel more complex paths. Due to their relatively short, 12 minute half-life, only high energy neutrons have a significant probability of reaching the Earth before decaying. The time-dependent flux of solar neutrons at a given distance from the Sun for a delta function production depends on the production spectrum at the Sun, the probability of neutron escape from the Sun, and the probability of neutron survival against decay in transit to the Earth. The escape probability is dependent on the neutron energy and production depth in the solar atmosphere so the probability of escape decreases with increasing depth. The probability of survival at a distance R is dependent on the velocity and therefore the energy of the neutron. In the case of the neutron observations from Mercury, a major advantage is

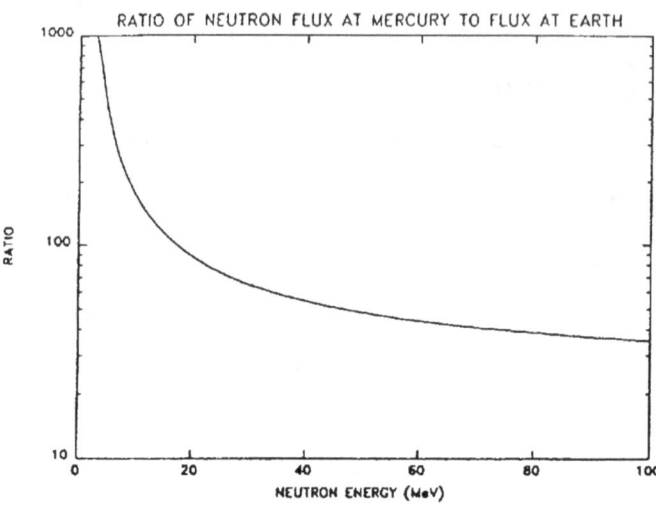

Figure 4. The neutron flux at Mercury relative to the flux at the Earth is shown versus neutron kinetic energy.

that the decay loss is greatly reduced for low energy neutrons in addition to a gain by the $1/r^2$ factor. For example, in a large flare observed by the Solar Maximum Mission (SMM), it was only possible to measure the transient neutron flux down to an energy of 50 MeV. The gain in flux at Mercury, compared to a near-Earth observation, is >10 for neutrons below 100 MeV and increases to more than 100 for neutrons below 10 MeV as shown in Figure 4. Clearly, the detection of low-energy neutrons, which is otherwise impossible, becomes quite practical with a Mercury orbiter.

As MeO approaches perihelion near 0.3 AU, the measurement of solar neutrons with an energy as low as 1 MeV is possible. The enhancement of both low-energy and high-energy neutron fluxes combined with the increased impulsiveness of neutron events will give new data not otherwise obtainable. At a distance of 0.3 AU the investigation of low energy neutron fluxes can, for example, determine if the Sun is a steady source of neutrons. The close proximity to the Sun will also allow improved correlation between neutron flux measurements and the stage of the event in which the neutrons arose. As discussed above, simultaneous observation of the gamma-rays and neutrons will allow the full properties of flares to be characterized for the first time.

Since neutrons and gamma-rays are so closely coupled to conditions in the flare region, yet are unaffected by magnetic fields, they provide the most direct information on the total number, energy spectrum, and angular distribution of the accelerated charged particles in the flare region. The measurement of solar neutron spectra over a larger energy range than was possible previously could give unique information on the primary proton spectrum and perhaps, given the shape of the neutron spectrum, allow the flare location to be inferred. It has also been suggested that a study of solar neutrons would provide a determination of the amount of ^{3}He in the solar atmosphere and the site of the nuclear reactions in the Sun. Sensitive neutron observations could also reveal the presence of quasi-continuous acceleration of ions at the Sun. Without a doubt, the study of solar flares at a distance of 0.3 AU could provide a new "window" into the processes of the Sun.

Hard X-Ray Directivity Measurements

Anisotropies in the velocity vector distribution of flare-generated energetic electrons can provide important clues about the acceleration and transport of electrons in the flaring region. The hard x-ray bremsstrahlung produced by a beam of electrons is preferentially forward, so that if an observation of the emissions is made at different angles from the beam direction, identical hard x-ray detectors would see a different response. For this reason, the flight of a hard x-ray spectrometer on a

Mercury orbiter and similar (cross-calibrated) detectors on spacecraft orbiting the Earth can provide vital information regarding the geometry of the accelerated electrons. The same detector needed for the gamma-ray line observations mentioned above can measure the hard x-ray spectral intensity. Highly eccentric orbits of the type favored for magnetospheric mapping will give a high-duty cycle. Thus, the likelihood that the complete flare, including low-energy neutrons, will be observable without interruption is higher than for low-Earth orbit (e.g., Solar Maximum Mission). Also, the background for a highly elliptical orbit may be lower.

Solar Energetic Particle Observations

A Mercury orbiter could also carry out unique solar energetic particle studies that cannot be accomplished by any other NASA mission except, briefly, by Solar Probe. The primary reason for this is that SEP measurements at 0.3 AU would allow the temporal history in the acceleration of a broad range of event types to be studied without the interference of interplanetary propagation effects that complicate the interpretation of 1 AU observations. The ambiguities facing observations at 1 AU are illustrated in Figure 5, which shows a series of curves calculated, using a standard numerical model of SEP propagation, based on a Fokker-Planck equation including diffusion, convection and adiabatic deceleration. In the model, the scattering mean free path is λ_r, and it is assumed to vary with heliocentric radius as r^b with particles at the Sun assumed to be injected as $\exp(-t/\sigma_T)$. The curves show values of λ_r, b, and σ_T that all give the same time of maximum at 1 AU. It is nearly hopeless to untangle the question of injection vs. interplanetary propagation with 1 AU measurements alone. In the very few events for which multi-spacecraft

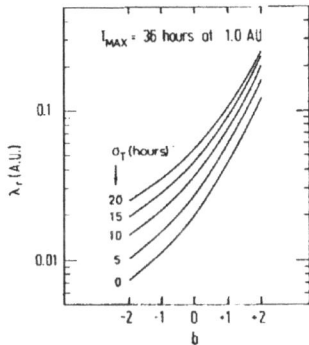

Figure 5. Parameter values for injection time constant σ_T, and interplanetary scattering mean free path λ_c, where $\lambda_c \propto r^b$, which give particle maximum arrival times at 1 AU that are all 36 hours after the flare at the Sun.

measurements reduce the number of free parameters, there is evidence that in large flares the injection of ~1 MeV particles takes place over periods of roughly 10 hours. The most convincing physical picture is that of large flares, where long injection time scales reflect acceleration at the Sun due to large shocks moving through the corona and accelerating particles out to several tens of solar radii. On MeO, located at 60 solar radii, this scenario could be easily and convincingly tested.

Another key solar particle event type to study on a Mercury orbiter would be the small, impulsive flares, which are often rich in gamma-rays, ^{3}He, and Fe. Figure 6 shows observations of such a particle event at 0.32 AU on Helios. The flux levels and anisotropies were extremely large, even though at 1 AU this event was modest in size. The dashed line in the figure shows the injection profile at the Sun based on a detailed numerical simulation. Note that fitting the particle data at Helios requires nearly a δ-function injection at the Sun. Thus, the particle data at 0.3 AU provides a tight constraint on the acceleration timescale. On a Mercury

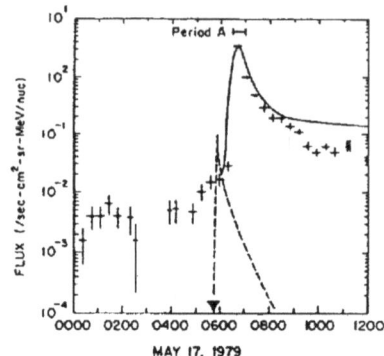

Figure 6. Helios-1 1.2 to 3.7 MeV/nucleon He fluxes at 0.32 AU for May 17, 1979. Shaded triangle marks time of flare at Sun; dashed lines show particle injection profile at Sun (arbitrary units); solid line is calculated to fit fluxes and anisotropies at Helios-1. The mean free path in this "scatter-free" event is $\lambda_r = 0.5$ AU.

orbiter many such events would be observed, and these observations, in conjunction with the gamma-ray and neutron studies would give critical insights into the processes operating in these small flares. The very fast time profile shown in the figure indicates that in this class of flares the acceleration is not via large-scale shocks, but is rather some rapid mechanism most likely confined to a small region.

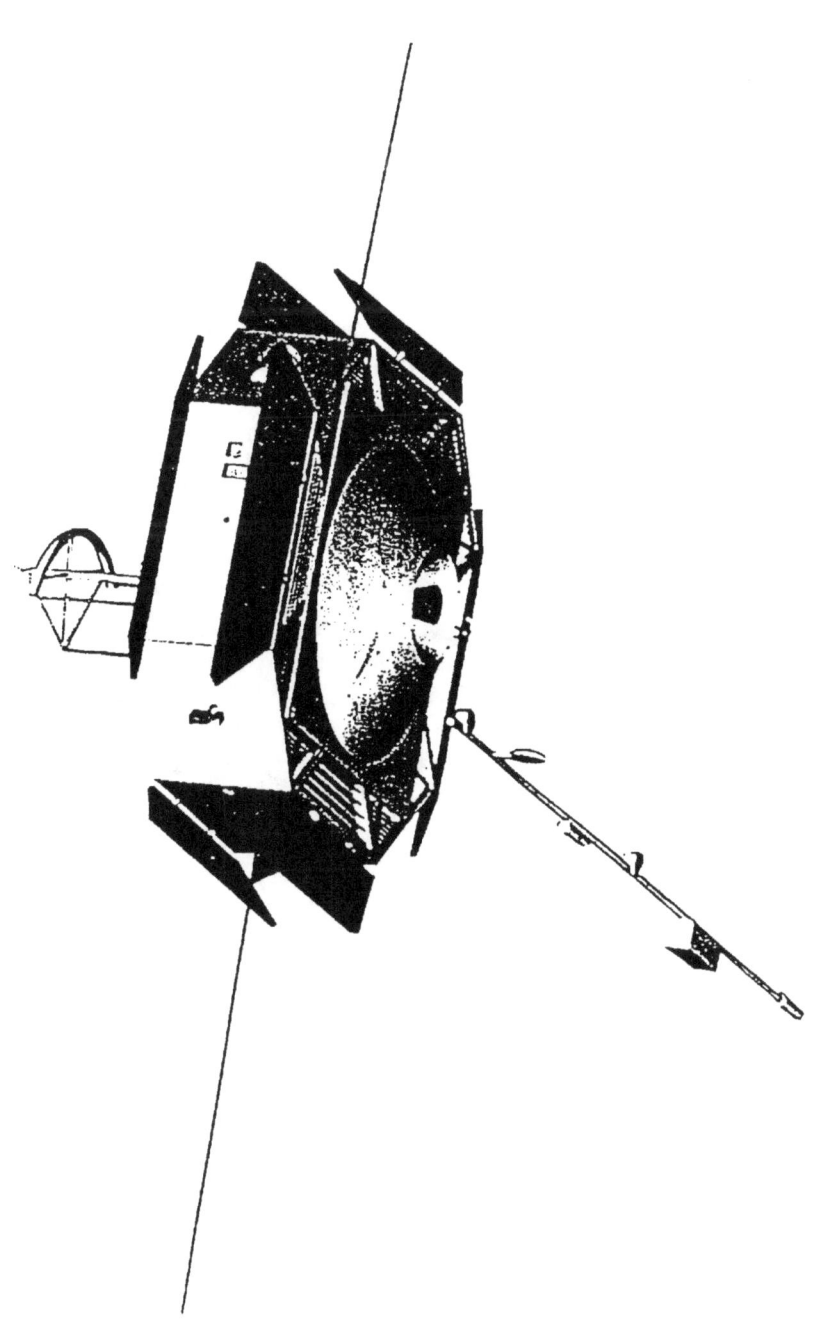

IV. MERCURY ORBITER MISSION

4.1 STRAWMAN INSTRUMENTS

4.1.1 Plasma and Energetic Particle Measurements

Magnetospheric plasma observations are critical to missions such as MeO because they contain signatures of energization processes, loss mechanisms, and particle sources for the resident particles. Particle distribution function measurements also provide measures of such physical effects as diffusion, transport, and loss operating in Mercury's exosphere and even the outer layers of the regolith. The low energy electron population is comprised of photoelectrons from the planetary surface and the ionosphere, thermal electrons arising from solar wind and magnetosphere sources, and suprathermal electrons from the interplanetary medium and magnetospheric acceleration sources. Various processes (e.g., transient inductive electric fields, field-aligned potential drops, and large-scale magnetospheric convection) can produce an intense energetic electron population from these low energy electrons. Consequently, observations of energetic electrons will reveal details of magnetospheric structure and dynamics.

Similarly, measurements of the ion distribution will tell us directly about the sources of magnetospheric plasmas, the dynamical processes responsible for energization and transport of these plasmas and indirectly about diffusion, transport, and loss processes in the Hermean exosphere and regolith. Measurements of the bulk ion distribution will give the overall plasma densities, temperatures, and flow velocities--measurements which will allow clear delineation of the plasma sheet, magnetotail lobes, magnetospheric boundary layers, and other physical regions. Ion composition measurements will help to identify exospheric and ionospheric sources and will help to distinguish these from solar wind plasma sources. Plasma composition also will aid in distinguishing among the proposed models of solar wind energy coupling and substorm particle acceleration.

Fast Plasma Analyzer

The fast electron analyzer (FEA) should make measurements in the 1 eV to 30 keV energy range as given in Table 1. This sensor should make efforts to compensate for spacecraft charging effects so that cold electron populations and photoelectrons from various sources can be well-characterized. Full three-dimensional measurements should be made as rapidly as possible. Several approaches are available; e.g., instruments with an intrinsic 4π scan, instruments using a 2π scan coupled with satellite spin, and multihead instruments. Whatever the approach, such instrumentation must be capable of measuring the three-dimensional distribution function in less than a satellite spin. The mass, power, and bit rate of a nominal FEA are estimated to be 4 kg, 5 W, and 10 kbps, respectively.

The fast ion analyzer (FIA) will make measurements, without mass or charge state discrimination, in the 1 eV/q to 30 keV/q range. Similar techniques are available for the FIA as were mentioned above for the FEA. It is envisioned that this sensor also will be able to make limited solar wind ion measurements, as well as characterizing ion densities, temperatures, and flows in the magnetosheath and plasma sheet at Mercury under conditions where a single ion species is known to dominate the total ion flux. With adequate telemetry, this class of measurement should also

Table 1. Plasma and Energetic Particle Instrument Characteristics

	Fast Electron Analyzer (FEA)	Fast Ion Analyzer (FIA)	Ion Composition Analyzer (ICA)	Energetic Electron Detector (EED)	Energetic Ion Detector (EID)	Solar Wind Electron Detector (SWED)	Solar Wind Ion Detector (SWID)
Measurement	Electrons	Ions	Ion Composition	Electrons	Ions	Electrons	Ions
Energy Range	1eV-30keV	1eV/q-30keV/q	5eV/q-50keV/q	10-500keV	10-500 keV/N	1eV-2keV	≤6 keV
Angular Coverage	3D	3D	3D	2D or 3D	3D	3D	Sunward Direction
Weight (kg)	4	4	10	15		4	6
Power (W)	5	5	12	15		4	6
Bit Rate (kbps)	10	10	10	10		2	2

be made on time scales of less than a spacecraft spin. The FIA requires 4 kg, 5 W, and 10 kbps in satellite resources.

Ion Composition Analyzer

The Ion Composition Analyzer (ICA) will make three-dimensional measurements over an energy/charge range of 5 eV/q to 50 keV/q. The sensor system may use either time-of-flight or magnetic selection techniques, but it should cover the mass range from H to Fe with good sensitivity for elemental determination. Time resolution, although not as critical as for the FIA, should still be an important design goal. Instrument mass is 10 kg with 12 W and 10 kbps as the power and telemetry rates, respectively.

Energetic Particle Detectors

As noted earlier, energetic particle measurements are critical to study basic plasma acceleration processes and to characterize large-scale magnetospheric morphologies. The Energetic Electron Detector (EED) system should cover the energy range from 10 keV to 500 keV. This gives reasonable overlap with the FEA and provides an improved geometric factor in the critical 10-30 keV range. To allow an assessment of open and/or closed field line geometries in the Hermean magnetotail on time scales compatible with substorm times, it is necessary to obtain a fast measurement of energetic electron fluxes from opposite directions. This can be accomplished by 4π detectors, 2π detectors plus a fast satellite spin, and/or multihead detector systems. Present imaging techniques applied to energetic particles will be of value for these measurements.

The Energetic Ion Detector (EID) will make ion measurements in the 10 to 500 keV/n range with good elemental determination. Such measurements can be achieved with time-of-flight (TOF) systems using 2π fan coverage plus spacecraft spin-scanning. As with the plasma sensors, this would give full 3-D coverage on the time scale of less than a spacecraft spin which should be adequate for most purposes. The EED and the EID together require 15 kg and 15 W. Within this allocation, it may also be possible to extend the electron energy range, at least in a portion of the sensor system, up to 1-2 MeV. This could allow measurements of relativistic electrons to identify Jovian and solar electron sources. The EED/EID system would require 10 kbps of telemetry for complete 3-D sampling.

If satellite telemetry is a limiting factor, the particle instruments described above can operate routinely in a survey mode at a relatively low bit rate (~1 kbps per instrument). In this mode, the data could be reduced to moments of the velocity distributions together with samples or averages of the total distributions. A low-duty cycle event capture mode will be employed on specified occasions whereby the full 10 kbps per instrument is buffered internally for later transmission to Earth.

Solar Wind Plasma Analyzer

During the cruise phase, and whenever the spacecraft is outside of Mercury's magnetosphere, opportunities will exist to take solar wind plasma measurements. Observations of the plasma electron and ion distributions in the inner heliosphere will provide critical information on the origin, acceleration, and evolution of the solar wind. While the energies of the magnetospheric and solar wind plasma populations overlap, experience has shown that the differences in the angular distributions, flux levels, and composition of the particles dictate the use of separate plasma analyzers. As shown in Table 1, the solar wind electron detector (SWED) and solar wind ion detector (SWID) will require 4 and 6 kg, respectively. The power and telemetry allocations are 4 W/2 kbps and 6 W/2 kbps for SWED and SWID. Close coordination between FEA/FIA and SWED/SWID will be required to ensure adequate observations of the boundary regions where energy transfer from the solar wind to the magnetosphere takes place.

4.1.2 Electric and Magnetic Field Instruments

It has been emphasized in the discussion of mission objectives that two important and unique characteristics of the Hermean magnetosphere are its small size and short time scales for convective processes. Both of these characteristics lead to the requirement of very high time resolution for the particles and fields measurements. The DC and AC electric and magnetic fields can be measured far more rapidly than most other plasma parameters, and will, therefore, provide a description of Hermean magnetospheric processes at small spatial and temporal scales. In the case of the electric field instrument, realization of the desired high time resolution requires two pairs of electric field probes in the spin plane. Triaxial fluxgate and vector helium magnetometers can meet the temporal resolution and sensitivity requirements for the magnetic field measurements. The inclusion of search coil magnetometers for measuring the magnetic component of plasma waves is highly desirable. In the terrestrial magnetosphere the electric field observations have proved vital for elucidating many important physical processes. No measurements of DC or very low frequency

Table 2. Electric and Magnetic Field Measurement Requirements

Quantity	Sensitivity	Range	Axes	Time Resolution	Sensor
DC magnetic field	< 0.1 nT	0-5000 nT	3	100 samples/sec	Triaxial fluxgate or vector helium magnetometer
AC magnetic field	3×10^{-6} nT/Hz$^{1/2}$	--	1-3	up to 30 kHz	Search coil magnetometer
DC electric field	<0.1 mV/m	$<0.1-10^3$ mV/m	2	up to ~20 kHz	Spherical double probe
AC electric field	<50 nV/m(Hz)$^{1/2}$	100 dB	2-3	up to ~800 kHz	Electric dipole*
Thermal electron density and temperature	$< 1\%$	$<1-100$/cm^{-3}	--	up to 2 kHz	Langmuir probe*
Density fluctuations	$< 1\%$	$\delta n/n$ of 1-50	--	up to 2 kHz	Langmuir probe*

*Note that the spherical double probes also function as Langmuir probes and can provide the spin plane dipole for the plasma wave instrument.

electric fields (< 10 Hz) have been obtained at any planetary magnetospheres except that of the Earth. Plasma wave measurements are also crucial to the realization of many primary heliospheric and magnetospheric science objectives. Such measurements will provide an opportunity to compare plasma and radio waves at a small, terrestrial-type planet to those of the large gas giants, as well as allow the examination of the role of wave-particle interactions in plasma heating and particle acceleration.

Table 2 lists the electric and magnetic field measurements needed to meet the goals of the Mercury Orbiter Mission. The following strawman

Table 3. Instrument Properties

Instrument	Weight (kg)	Power (W)	S/C Requirements	Telemetry** rate (kb/s)
DC Magnetometer		4	Magnetically clean	5
Sensors (2 sets)	2			
Electronics	3.3			
DC Electric Field/ Langmuir probe		7	Conducting S/C; Electrostatically clean; some real-time command capability	10
Boom & sensors (2-axis)	13.2			
Electronics	5			
Plasma/Radio Waves Receiver*		10	Electromagnetically clean	10
Search coil (3 axis)	1.2			
Electronics	6.0			

*Utilizes spin plane DC booms
**All instruments require 1-2 kb/s for low resolution survey modes

instrument complement has been devised for obtaining the E and B fields data: (1) a triaxial magnetometer (either fluxgate or vector helium) mounted on a boom, (2) a two-axis, double probe electric field instrument, and (3) a plasma wave receiver utilizing the double probe spin plane electric field booms, a triaxial search coil magnetometer, and, possibly, a short dipole along the spin axis. The weight, power, and telemetry rate requirements for these instruments are presented in Table 3. Note that we assume a spinning spacecraft. For electric and magnetic fields measurements, these spacecraft provide distinct advantages over three axis oriented designs by virtue of their ability to deploy very long booms perpendicular to the spin axis and the availability of spin modulation techniques for determining instrument offsets and direction-finding.

Vector Magnetometer

The scientific objectives to be addressed by MeO require clean and unambiguous magnetic field measurements. Spacecraft generated DC fields and noise should be below the instrument sensitivity levels to fully exploit the capabilities offered by present-day measurement technologies. Either of the two currently used techniques, triaxial fluxgate and vector helium magnetometers can operate reliably in the Mercury environment. The experience gained on missions such as Helios, Pioneer, and Voyager guarantees the feasibility of a long duration mission to Mercury. Based on the results of the Helios program, thermoelectric currents should not pose major problems for the magnetometer on MeO.

The proper operation of the magnetometer requires a magnetic cleanliness program to minimize spurious fields. Several in-flight techniques have been devised to check or to determine the spurious magnetic fields close to the sensor. A well established approach is that of using two identical magnetometers on a single, long boom to eliminate spacecraft fields during data processing on the ground. A spinning spacecraft allows for the easy removal of spurious magnetic field components in the spin plane through the detection and removal of the resultant spin modulation.

The need for very high sampling rates in some regions can be reconciled with the modest data rates through use of a burst memory. The requirement for accurate, evenly spaced, temporal sampling is dictated by the need of preserving the spectral integrity of the collected data (i.e., the sampling must then be time-synchronous). A number of preprogrammed or command-programmable on-board computations should be foreseen, having been already been done or being included on other missions (e.g., snapshots of special events, computation of power spectra or structure functions, Fast Fourier Transforms (FFTs) of data, averages and variances over appropriate time intervals, etc.).

Vector Electric Field Instrument

Electric field measurements in the Earth's magnetosphere have been crucial to our understanding of processes which energize and scatter particles. At Mercury, many similar phenomena are expected to occur and electric field measurements will be even more vital due to their smaller scale sizes and more rapid temporal evolution. The electric field measurement requirements can be met by a double probe electric field instrument which consists of two orthogonal pairs of wire booms. Each boom is a centripetally deployed wire, nominally 50 m long, with a spherical sensor at the end. The electric field measurement is made by measuring the potential difference between the probes at the ends of opposing booms. The spherical sensors can also be operated as current collecting Langmuir probes to measure the thermal electron density and temperature. By sweeping the sphere voltage, the current-voltage characteristics are obtained. Density fluctuation measurements for studies of electrostatic waves and structures such as double layers are obtained by operating the probe as a current collector at a fixed voltage. The large dynamic range of the electric field instrument is necessary to resolve the large-scale convection field, MHD wave fields, low frequency waves such as lower hybrid waves and electrostatic ion cyclotron waves, and structures such as electrostatic shocks, double layers, and spiky electric fields which have been shown to provide the particle acceleration in several regions in the terrestrial magnetosphere such as the auroral zone and bow shock. Finally, the electric field instrument also monitors the spacecraft potential.

Throughout the entire Mercury orbiter mission, the double probe instrument will be operating in a low density environment which, for this application, is defined as a density such that the photoemission current dominates the electron thermal current. The sphere potential in the electric field mode is determined by the overall current balance. When the thermal current is small, the instrument sends a current to the sphere to balance the photoemission. This current is determined by microprocessor control to minimize the sheath resistance and, therefore, to increase the accuracy of the measurement. In this low density regime, the electron temperature is obtained, not from the slope of the I-V curve, but rather from focussing effects which are observed at very high positive potentials. This technique for measuring DC and low frequency electric fields in a low density plasma has been flown successfully on the ISEE and CRRES satellites and was further refined in designs under development for CLUSTER and POLAR.

Plasma/Radio Wave Science Instrument

The investigation of magnetospheric plasma waves and the wave phenomena associated with the solar wind-magnetosphere interaction requires coverage of the frequency range from the sodium ion cyclotron frequency (about 0.1 Hz) to well above the solar wind plasma frequency (of order 100 kHz). Accommodation of solar radio burst measurement requirements could extend the frequency range of the plasma/radio wave science (PRWS) instrument to 1 MHz or higher, depending on the science to be performed relating to the solar emissions. Both electric and magnetic components of waves need to be measured to ensure that electromagnetic and electrostatic wave modes are correctly identified. However, since the maximum electron cyclotron frequency in the magnetosphere is about 15 kHz, the magnetic measurements can be limited to below this value as higher frequency modes must be electromagnetic in character. It can be assumed that the electromagnetic/electrostatic nature of solar wind plasma waves is well enough understood from near-Earth satellites that there should be no problem in identifying the solar wind emissions as electrostatic or electromagnetic. The study of shock turbulence, however, might benefit from higher frequency magnetic measurements.

The minimal requirements for the plasma/radio wave instrument are a single, 10 m tip-to-tip electric dipole antenna and a single-axis magnetic search coil. Sensitivity can be enhanced by extending the dipole axis to greater lengths, perhaps by making use of the vector electric field instruments' double probe booms. Additional information on the wave mode can be obtained by the use of triaxial search coil antennas and the inclusion of a short (up to 10 m) dipole extended parallel to the spacecraft spin axis.

The PRWS instrument would benefit by being very flexible in the way signals are analyzed on board to make optimal use of the available data rate. Sweep frequency receivers could be utilized to characterize the wave spectrum at low temporal resolution, but with moderate frequency resolution. Multichannel receivers could provide important measurements for understanding the role of plasma waves in the substorm process.

Ultimately, one would like to have very high time resolution waveform measurements up to 15 kHz, and possibly at higher frequencies by mixing with an oscillator, for at least brief intervals. Such a capability can easily utilize 100 kbps or more of data rate; hence, the need for flexibility in the implementation to make maximum use of the available downlink. This flexibility should include a burst mode triggered by onboard monitors for events such as substorm onset signatures, ground commands, and stored sequence commands. The waveform should also have low data rate modes which capture waveform samples at very poor temporal resolution to enable surveys at high spectral resolution. The overall data rate of the instrument should average at least 3 kbps with higher rates for telemetering the burst mode.

Summary

There are numerous design tradeoffs and resource allocation decisions that will need to be addressed in more detail when further studies of the Mercury Orbiter Mission are made. These include the length and number of the instrument booms, the number of magnetic search coils, the number of vector magnetometers, and the technical feasibility of including an electric dipole parallel to the spacecraft spin axis. In addition the MeO SWT did not specifically consider electron beam electric field instruments. Three reasons for this decision are: (1) their frequency is limited to ~100 Hz and it would, therefore, still be necessary to include dipoles for the PRWs measurements, (2) it would not be possible to obtain thermal electron density, density fluctuations, or temperature measurements without adding a Langmuir probe which would also require some additional spacecraft resources, and (3) the beam instrument may not be able to make measurements in the highly turbulent environment which is expected at Mercury. However, these concerns should be re-examined at some future date when this technique has seen application in magnetospheric missions at Earth (i.e., CLUSTER).

4.1.3 Surface Imager

The strawman camera system described here for the Mercury Orbiter Mission is designed to take panoramic pictures from a spinning spacecraft. Using a CCD line array, the spacecraft spin performs the panoramic scan. Many of the components of the camera are already in development for current missions such as the narrow-angle Mars Observer Camera (MOC). Thus, their availability and performance are assured.

The strawman camera has a 50 mm focal length, f/4 optical system to focus an image on the 2048 x 1 pixel charge coupled device (CCD), as shown in Figure 7. This system provides an instantaneous field of view of 0.26 mrad and a half-angle field of view of 15 degrees.

The line array is mounted parallel to the spacecraft spin axis. An external shutter is required to protect the CCD from imaging the Sun. The camera will take pictures 2048 pixels wide with lengths determined by the programmable scan of the spinning spacecraft. The CCD is clocked to give square pixels during exposure. During readout, the charge passes through an analog-to-digital converter, then to the central processing unit (CPU) for data compression and then to the mass storage unit to await transmission to Earth.

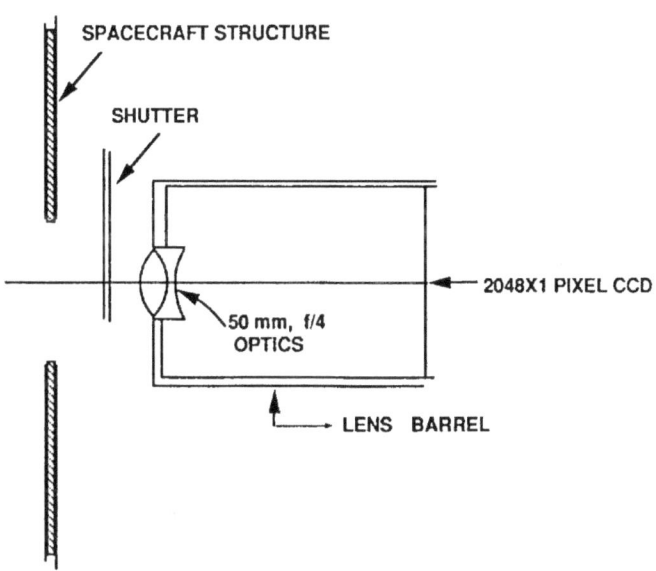

Figure 7. Schematic of a Mercury Orbiter imager.

The 2048 x 1 pixel CCD has 13 micron pixels and is produced by Ford Aerospace Corporation for the MOC. The operating temperature is about -20°C; a cold finger and radiator might be required to maintain this low temperature. If the radiation environment is severe, some shielding might be needed. The 50 mm focal length and the 10 RPM spacecraft spin rate give a readout line time of 0.25 millisecs. This corresponds to an 8.25 million pixels/sec readout rate; two 8-bit A/D converters are required, as the maximum rate for each is 5 million pixels/sec. The CPU is a National Semiconductor Series 3200, radiation hardened 32-bit microprocessor. A number of data comparison algorithms have been tested and a compression factor of 2 to 4 is readily available. Higher compression factors might be useful for this mission. A 12 megabyte buffer is included so the spacecraft communication data rate can be matched. With a 10 kilobit communication data rate, the buffer can be read out in about three hours. The camera characteristics are summarized in Table 4.

Table 4. Line-Scan Panoramic Camera Characteristics

Optics	
Focal length	50 mm
Focal ratio	f/4
Spectral bandpass	500nm - 900 nm
Detector	
2048 x 1 pixel line array CCD, 13 micron pixels	
Line time	0.25 millisec.
Readout noise	Approx. 200 e-
Dark current noise	15 e- equivalent
Operating temperature	-20°C to -50°C
Mass Storage	12 megabytes
Pixel Output Rate	8.25 M pixels/sec
Data Compression Factor	2 - 4
CPU	Rad-Hard 32-bit microprocessor
Mass	4 - 6 kg

The strawman MeO mission design endorsed by the SWT and described in later sections involves two spinning spacecraft. Both spacecraft will be placed in 12 hour polar orbits, as shown in Figures 8a and 8b, one with periapsis above the equator and another above the north pole. Periapsis altitude is 200 km for both orbits. The two orbital plane inclinations are separated by about 60 degrees if the strawman mission scenario to be described in later sections is followed. The nominal spin axis orientation of both spacecraft is northward.

Mercury has an orbital period of about 88 days and a rotational period of 58.6 days with its spin axis normal to its orbital plane. With these orbital characteristics, all surfaces of Mercury are eventually illuminated by the Sun and most areas are available for imaging. Due to the 3:2 spin-orbit coupling, all areas can be imaged at least three times per spacecraft during the nominal two Mercury-year long final phase of the mission

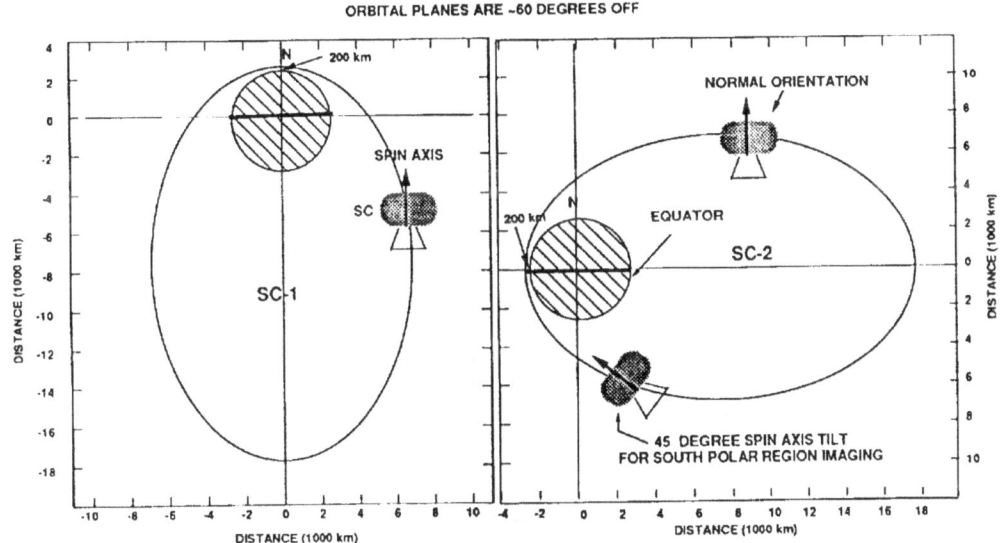

Figure 8a. Schematic of SC-1 polar orbit with spin axis pointing north.

Figure 8b. Schematic of SC-2 polar orbit with spin axis tilted 45° for south pole region imaging.

depicted in Figure 8. The solar illumination angles of these images are about 4 hours apart because of the spin-orbit coupling and the non-precessing nature of the polar orbit. The deviations from the exact, 4 hour separation in phase angles of three images are due to the eccentricity of the Mercury orbit. Observing surface details at many phase angles is very important to the interpretation of the topography. The imaging of the planet using dual orbiters greatly enhances the quality of the imaging in this respect.

For the polar orbiter with periapsis at the north pole, the imaging interval lasts 40 minutes per orbit. The orbital period is 23 minutes for the orbiter with its periapsis at the equator. For each spacecraft the camera will point to nadir only at the equator. At other latitudes, non-

Table 5. Imaging Characteristics of MeO

SC-1 (Periapsis at N-pole)			SC-2 (Periapsis at equator)**		
Latitude	Line of	Pixel*	Latitude	Line of	Pixel*
S/C Image	Sight Alt	Length	S/C Image	Sight Alt	Length
(deg)	(km)	(m)	(deg)	(km)	(m)
-20 -60	4538	2326	0 0	200	52
-10 -22	2926	820	10 11	221	59
0 0	2152	560	20 22	289	81
10 17	1672	454	30 35	426	135
20 31	1351	409	40 51	696	285
30 43	1130	404	50 78	1483	1820
40 55	988	449	51 89	1929	30622
50 67	930	619			
60 84	1125	2597			

* Central pixel with foreshortening
** Same for north and south

nadir pointing degrades the resolution of the images. SC-2 will provide equal north and south coverage with high resolutions near the equator. SC-1 will provide good coverage of the high northern latitudes. The spacecraft latitude, the imaged latitude along the line of sight of the camera, the imaging range, and the typical pixel resolution are given in Table 5.

The overall imaging resolution (line pair) provided by each spacecraft is shown in Figure 9. Each curve provides the variations in latitudes and resolutions of imaging along the CCD array when the spacecraft is at a particular latitude.

The image resolution near the poles is poor due to the non-nadir pointing geometry. The steepness of the curves near the poles indicates that for a given scan, the resolution varies dramatically along the length

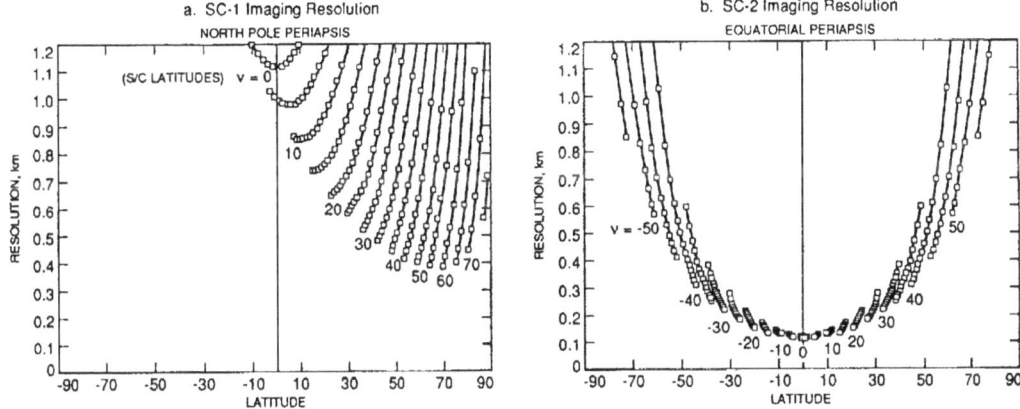

Figure 9. Imaging resolutions vs. latitudes.

of the array. The best resolution will be obtained in the limb pictures where the look angle to the surface is 15°. The north pole is imaged by SC-1 with a resolution of 640 m. SC-2 coverage with (line pair) resolution of better than 1 km is possible up to ± 77°. To attain more satisfactory coverage of the polar regions, the MeO spacecraft can be re-oriented for short periods to attain more nadir-oriented viewing of the polar regions. For example, about every 44 days (one-half of Mercury's 88 day orbital period) the plane of the orbit will lie in the plane of Mercury's terminator and the thermal input from Mercury presents no threat to the spacecraft. At these times, it is possible to tilt the spacecraft spin axis in the plane of the orbit to achieve advantageous viewing, as shown for SC-2 in Figure 8b. Using this technique, the coverage of the south polar region will be much improved, albeit not as good as the SC-1 imaging of the north pole, because of the greater distances between SC-2 and the south pole. At a distance of about 2150 km, the pixel size at the south pole is 660 x 800 m.

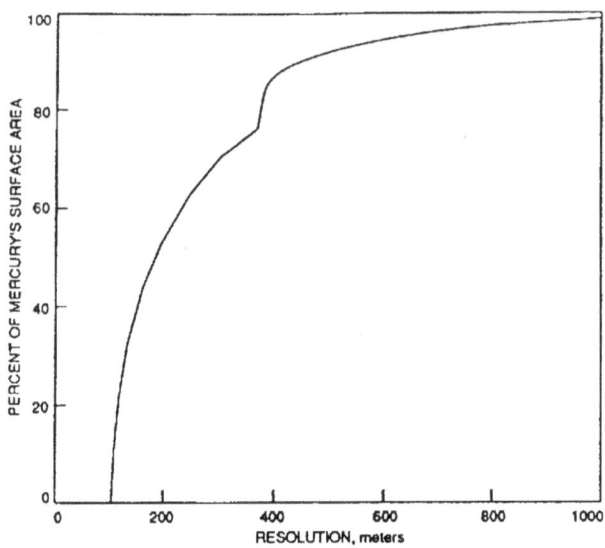

Figure 10. Cumulative imaging coverage.

It is desirable to obtain full coverage at resolution of 1 km. This is possible except for a small area south of 80° latitude. Figure 10 shows the percentage coverage of the planet at various resolutions with the nominal spacecraft attitude. About 99% coverage at 1 km or better resolution is possible and 60% coverage at better than 250 m resolution is expected.

4.1.4 The Geochemistry Experiment

The planet Mercury has become an object of ever-growing scientific interest over the last decade. As discussed earlier, the recognition has grown that close study of this planet is essential in order to address a variety of fundamental scientific questions.

A geochemical experiment is essential for MeO because of its ability to provide:

1) estimates of global concentrations of key elements (Al, Mg, Si, Fe, Ti, K, Na, Ca) that would indicate which, if any, of the paradigms proposed to explain the origin of the Solar System is most accurate, and would allow for much greater understanding of the early Solar System environment.

2) estimates of variations in concentrations of these elements for major terranes, to indicate the nature and extent of geochemical differentiation of the interior.

3) estimates of variations in concentrations of these elements (excluding Na and Ca) for major geological features, to determine the history of Mercury's surface, and, when combined with image and relative age data from the imaging experiment, the times and extent of volcano-tectonic episodes on Mercury.

4) estimates of certain particle components of the solar wind to determine, when combined with the results of other solar wind experiments, the nature of the solar wind and its interaction with Mercury's surface.

X-Ray Fluorescence Experiment

The strawman X-Ray Fluorescence (XRF) detector system consists of four gas filled proportional counters mounted in the configuration shown in Figure 11a with differential filters. One of the detectors will act as a solar monitor. This detector will have a pinhole window to prevent saturation by the solar flux. The other three detectors will be optimized for the detection of Si, Al, and Mg, respectively. All three detectors will be used differentially to detect Na and Ca. A 100 channel recorder,

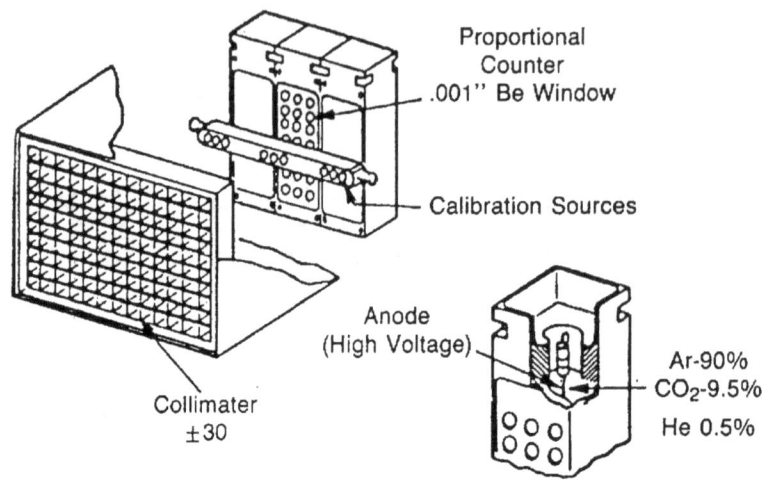

Functional configuration of the X-ray spectrometer.

Figure 11a. X-Ray spectrometer.

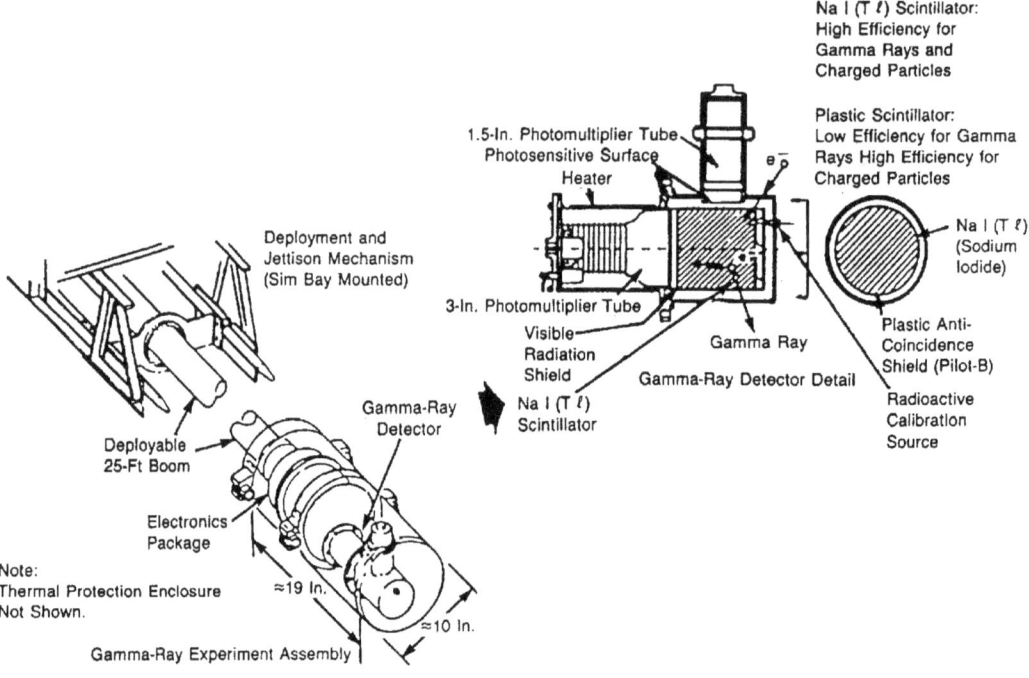

Figure 11b. Gamma-ray spectrometer.

instead of the eight channel recorder used for the Apollo mission, is recommended to record the count rates as a function of energy. The range of the detector system will probably be somewhat expanded, to cover the .5 to 2.75 KeV range.

At Mercury, the solar flux is an order of magnitude higher than at the Moon. As a result, the XRF experiment has some additional thermal protection requirements. The detectors must never view the Sun directly, to avoid the saturation which would result in a long recovery period and would affect the measurement capability of the instrument, and to prevent the shortening of the instrument's lifetime. The detector system, as configured on the Apollo missions, would admit one Watt per square centimeter to the spacecraft (a total of 75 Watts) and would thereby heat the detectors to an unacceptable temperature. To meet additional thermal protection requirements, we require template-style shuttering for the experiment, with the shutter opening only during ≤ 1 second intervals around nadir to collect data. In addition, we require that an additional 1 mil equivalent of highly reflective beryllium be placed over the detector windows. Previously, the detector windows were covered by only a 1 mil equivalent of beryllium; now, the windows would be covered by 2 mil equivalents. Greater thicknesses of beryllium would too greatly attenuate the signal. We also require that gold-flashed Mylar instead of aluminized Kapton be used in thermal blankets associated with the XRF experiment, to prevent interference in the detection of aluminum from the surface.

Additional careful thermal studies must be done to determine whether the XRF experiment can survive under the proposed conditions, even with additional thermal protection as proposed above. The problem is that the effective window size for the proposed proportional counter detectors is relatively large, and may heat the interior gas to unacceptable limits (above 40°C).

A possibility exists that a solid state HgI detector system now under development might be available in a few years to replace the proportional counters. The use of this new detector would result in the elimination of the thermal problem due to the much smaller area of the detector window and the ease of incorporating thermal electric cooling options. A mass savings of 2 to 3 kg would also result from the use of a solid state detector system.

Gamma-Ray Spectrometer Experiment

The Gamma-Ray Spectrometer (GRS) detector system shown in Figure 11b consists of a solid state NaI detector, similar to the one flown on the Apollo 15 and 16 missions. However, in this case shielding will be required around the detector, to eliminate the sky background. The BiGe shield will allow determination of direction of the source of detected gamma-rays and, thus, allow elimination of sky sources and effective collimation of the signal. Generally, the amount of shielding required and the spacecraft spin rate are correlated. The necessity of shielding, which adds mass, and the possibility of thermal problems preclude the boom mounting of the instrument. Aluminized kapton blankets may be used for the GRS instrument, provided that no part of the blanket is ever within the XRF instrument field of view. The necessary addition of a shield (for spin rates up to 5 or 10 RPM) will add 10 kg to the GRS experiment, making the total weight requirement for the GRS package 25 kg. The potential availability of a new generation of lighter weight photomultipliers (microchannel plates) could reduce the weight of the instrument by up to 2 kg.

Additional navigational requirements for the geochemistry experiment package are necessitated due to the spinning nature of the spacecraft. In order to acquire and optimize the signal from the surface for short time bursts near nadir, the following pieces of information are required in real time with an accuracy on the order of 10 msec: solar position (via Sun sensor), spin rate, and nadir position. These data would be used to operate the shutter on the XRF spectrometer, and to control data accumulation operations for both detector systems.

Experimental Capabilities

The geochemistry experiment package is capable of providing concentration data for the elements Al, Mg, Si, Na, and Ca from the XRF experiment, and Fe, Ti, K, Th (and/or other elements which are the products of natural radioactive decay) from the GRS experiment. These data could potentially be provided in one or more modes as shown in Figure 12:

a) Local Mapping (under 50 km)

As the data indicate, mapping at this scale, similar to mapping of Al, Mg, and K variations for approximately 10% of the lunar surface, is not possible here.

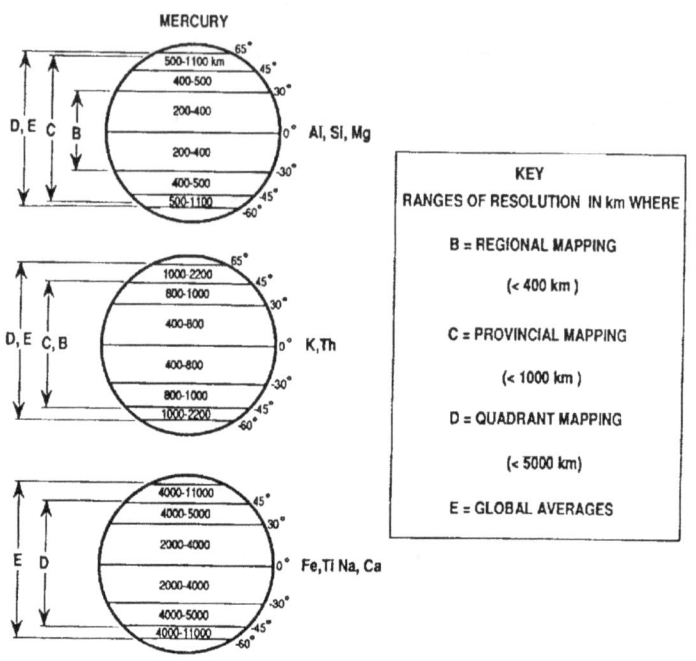

Figure 12. Estimated spatial resolutions at 20° collimation.

b) Regional Mapping (under 400 km)

As the figure illustrates, this type of coverage is available for Al, Mg, and Si data, and to a minimal extent, for K and Th data in the equatorial region. This type of mapping, which can show larger geological features quite clearly, was performed for Fe and Ti for 20% of the lunar surface. Data are available up to 30 degrees from the equator for Al, Mg, and Si data. Major volcanic or impact depositional features, such as plains deposits, should be detectable with these data. It is crucial that data with this resolution be provided for as many elements as possible. At spin rates of 10 RPM, data will have lower signal-to-noise ratios, and losses of data--especially at higher latitudes--will occur, inevitably lowering actual spatial resolution.

c) Provincial Mapping (under 1000 km)

This type of coverage is available for Al, Mg, Si, up to 55 degrees from the equator, and for K and Th up to 45 degrees from the equator. Once again, coverage could be limited for K and Th at higher spin rates due to poorer statistics. Major geochemical terranes should be discernible from these data.

d) Quadrant Mapping (under 5000 km)

This type of coverage is at least partially available for all data. It is available below 65 degrees latitude for K and Th, and below 30 degrees for Fe, Ti, and possibly Ca and Na. At 10 RPM, poorer statistics may be a problem for the data with already marginal capability for this resolution, particularly Na and Ca. Al and Mg can be mapped up to 65 degrees latitude at this resolution, if one chooses to average. Major changes in the nature of surface material within hemispheres, as well as global asymmetries should be discernible from these data.

e) Global Averages

Global averages of all elements considered here should be producible from these data. However, for higher spin rates, estimates will be less certain and will be much more heavily influenced by the chemistry of the equatorial region, with the poles having very little weight.

At 20 degrees collimation, the experiment flown on the equatorial periapsis orbit spacecraft (1) contributes the bulk of the data, with enhancement from the experiment on the polar periapsis orbit spacecraft (2). With the use of deconvolution techniques, the bulk of the mapping data could be improved in quality (signal-to-noise ratio) by including data from the second spacecraft. Also, the quality of global average data would be enhanced, due to improvement in signal-to-noise ratio by additional data averaging. Spacecraft 2 contributes data with comparable or better resolution between 60 and 65 degrees latitude in the northern hemisphere, until the planet no longer fills the field of view.

Summary

A geochemical experiment package composed of XRF and GRS instrumentation must be considered as an essential part of any future mission to Mercury. No other approaches presently exist to provide these data. Without them, spectral and color difference data will remain largely uninterpretable. Such data for Mercury, including Fe, Ti, Na, Ca, Al, Si, and Mg abundances, are crucial for understanding the early Solar System. Some geochemical data on the scale of lunar data must be acquired to map the geochemistry of major geological features for at least a portion of the surface, thereby producing data which is convincing to planetologists.

4.1.5 X-Ray, Gamma-Ray and Neutron Instrumentation for Solar Physics

In addition to the core magnetospheric and planetological instruments described above, the Mercury Orbiter Mission offers a unique opportunity to obtain other important scientific measurements of solar photons and particles.

A small NaI spectrometer (7.6 cm x 7.6 cm) can make extensive observations of solar flares from Mercury during the maximum of cycle 23 (circa 2002). Because of the $1/r^2$ factor, this small instrument would have a sensitivity roughly equivalent to that of the SMM spectrometer which used

7 crystals (7.6 cm x 7.6 cm). During the maximum, the detection rate for gamma-ray flares with emissions above 300 keV would be expected to be ~5 per month. This spectrometer need not be boom-mounted. Both flare bremsstrahlung continuum and nuclear lines (especially 2.223 MeV) can be studied up to a few MeV, and be correlated with neutron observations. This instrument should be cross-calibrated with similar instruments expected to be on other space probes, so that maximum information will be obtained on flare x-ray anisotropy.

Priority should be placed on good time resolution. Time structure in gamma-ray flares is known to be ~1 s or less, but the time resolution of spectra from the SMM detector was 16 s. Note that an omnidirectional detector would also be valuable for gamma-ray burst measurements, giving a long baseline for the time-of-flight technique when used in conjunction with Earth orbiting detectors.

The x-ray/gamma-ray spectrometer used for planetary studies could also be used for the solar flare studies, if the spacecraft is reoriented for long-term observations of the Sun. Alternatively, a neutron detector which is also sensitive to gamma-rays (e.g., a scintillation detector) could be used, but sufficient energy resolution for spectroscopy would be necessary. Barring this, a dedicated (omnidirectional) solar instrument is needed, which is continuously exposed to the Sun, except for planetary occultation.

Because decay loss for MeV neutrons remains significant even at Mercury, it is necessary to measure a transient neutron flux of magnitude 10^{-2} neutrons cm^{-2} s^{-1}. Since production of neutrons in the spacecraft can exceed this level, the neutron detector should be boom-mounted at a distance of at least 5 m from the body of the spacecraft. The most sensitive MeV neutron measurements can be made when the orbiter is at an altitude of several planetary radii, where the surface leakage flux is reduced below that from the spacecraft itself. The required instrumentation must effectively detect MeV neutrons in the presence of a strong, hard x-ray and gamma-ray flux from the flare.

Since the solar neutron flux below 100 MeV at Mercury is more than an order of magnitude larger than at Earth, even a rather small detector can surpass the response of the spectrometer on SMM. For example, a proton recoil detector of organic scintillator with a threshold at 1 MeV would have an effective area equal to or greater than that of SMM at all energies. This would allow the measurement of approximately two to five neutron events during a 1-year period near solar maximum. The same neutron detector (which is omnidirectional) can be used to measure the Mercury leakage flux, especially if the orbit is elliptical, since the variation in the detector's rate versus altitude can be used to separate out the spacecraft background. The total detector mass and power are estimated to be 20 kg and 10 W and the peak data rate will be 0.5 kbps.

Optimization of the x-ray, gamma-ray and neutron measurements, as well as trade-off considerations to minimize weight and power, require the following studies:

o Evaluation of neutron detectors, such as organic scintillators, fission chambers, He^3 proportional counters, Li^6I scintillators, etc.

o Combining the gamma-ray and neutron and solar energetic particle detectors.

o Model calculations of gamma-ray background.

o Model calculations of spacecraft neutron background versus boom length.

o Model calculations of the Mercury neutron leakage flux spectrum versus altitude.

4.2 SPACECRAFT DESIGN

The objective of this study was the conceptual design and characterization of a spacecraft that meets the fundamental project, science, and mission needs. Engineers (see Appendix A) from nine technical disciplines made contributions to this ≈ 1.5 workyear design study. Effort was not expended on meeting low-priority desires that impose significant penalties on spacecraft operations, performance, resources, reliability, or cost. These will be addressed later in Phase A engineering studies.

A system design method was employed that in each iteration included consideration of all subsystems but, at any given time, focused on the most serious unresolved system-level challenges. Emphasized in the spacecraft design were simplicity and efficiency, adequate capability, and minimized overall and early-year costs. Two spacecraft are baselined for the MeO mission to meet science requirements for comprehensive sampling of

Mercury's magnetosphere and surface and to provide redundancy in the event a single spacecraft were to fail.

Design Drivers/Considerations

Launch Vehicle--A single, Titan IV/Centaur with a solid rocket motor upgrade and a 56-foot shroud is used for the simultaneous launch of the two spacecraft.

Commonality--Two identical spacecraft are required. This simplifies design, test, and operational phases, reduces cost, and expands the launch contingency options; i.e., the launch stack order can be reversed and either spacecraft can be launched alone.

Mass--Mission mass allocations are 200 kg for launch adapters, 800 kg for each dry spacecraft, and 1600 kg for the total propellant mass in each spacecraft. (Mass allocations assume a main engine specific impulse $\geq$ 308 s.)

Trajectory/Orbits--The solar range during the mission drops from 1 AU at Earth to as low as 0.307 AU at Mercury perihelion. The Mercury orbit period is $\geq$ 12 hr; periapsis altitude is $\geq$ 200 km; and apoapsis altitude is between 15,200 and 200,000 km. Occultation periods are constrained to be $\leq$ 2.2 hr.

Science Payload--Basic characteristics of the strawman instruments, including their instantaneous fields of view, are listed in Table 6.

Table 6
Strawman Instruments

	SEN. LO.[1]	FOV (°x°)[2]	RATE (KBPS)	LOAD (W)[3]	MASS (KG)[4]
DC Electric Field Analyzer	WB	---------	.064-10	7.0	18.2[5]
Energetic Particle Detector	P	12x180 50x180[6]	1-10	15.0	15.0
Fast Electron Analyzer[7]	P	15x180[6]	1-10	5.0	4.0
Fast Ion Analyzer[7]	P	15x180[6]	1-10	5.0	4.0
Gamma/X-Ray Spectrometer	I	±10/±20	1.2-2.4	14.3	17.0[8]
Ion Composition Plasma Analyzer[7]	P	15x180[6]	1-10	12.0	10.0
Solar Wind Analyzer	P	45x180 70x180 160x180	0.4-4	10.0	10.0
Line-Scan Imaging (and TEC)	I	0.015x30	10	11.0	5.1
Magnetometer[9]	SB	---------	1-5	4.0	5.3
Radio/Plasma Wave Analyzer	WB/SB	---------	.032-10[10]	10.0	7.2
Solar Neutron Analyzer	SB	---------	0.5	10.0	10.[11]

1. Sensor locations are: Internal, Perimeter, Science Boom, Wire Boom.
2. Fields of View are expressed either as Clock°xCone° where the angles are with respect to the spin axis (or line parallel to the spin axis) or as ±X where X is the angle from the instrument boresight.
3. Loads are shown for the normal operating mode.
4. Blanket and Sun shutter masses are not included in this table.
5. The DC EFA mass includes 3 kg for wire boom rewind, position readout, and (emergency) cutters.
6. Requested "cone angle" FOVs for the EPD, FEA, FIA, & ICPA were 360° (i.e., ± 180°). Instead they have been provided 180° FOVs (which still provide full-sky coverage in one spacecraft revolution) at twice the rotation rate (which helps offset any loss in time resolution). This decreases the heat load on the sensor by 2X, greatly simplifies the interface design, and decreases solar panel EMI.
7. The FEA, FIA, & ICPA share a common sensor package, processor, and electronics housing.
8. It may be necessary to add an active radiation shield to the GRS, increasing its mass by $\leq$ 10 kg.
9. Calibration coil mass of 0.5 kg is not included in this table (but is included in the spacecraft mass).
10. It has been assumed that a buffer will be added to the PRWA to store data from its 300 kbps "burst" mode. The maximum instrument data rate has been increased from 1 kbps to 10 kbps to enable reasonably fast buffer readout.
11. The requested SNA mass was 17 kg, however, the allocation from the SWT is 10 kg.

Thermal Environment--The direct solar flux on the spacecraft ranges from zero during occultations to 1 solar constant (SC) at Earth, to 10.6 SC at Mercury perihelion. Solar flux reflected from Mercury varies from zero in occultation to a maximum, usually < 1 SC, when the spacecraft is above the illuminated surface. Total solar flux depends on solar range, phase angle, altitude, and surface albedo in the vicinity of the spacecraft. Heat flux on the spacecraft reradiated from Mercury (in terms of equivalent solar constants) ranges from zero when the spacecraft is at apoapsis and in occultation to $\approx$ 8.5 SC when the spacecraft is at periapsis above the subsolar point and Mercury is at perihelion. Angular separation between direct solar flux and heat flux from Mercury is unlimited, and while the Sun is small in the sky, Mercury reaches a large angular extent. Also, most of the energy from the Sun is at short wavelengths whereas that radiated by Mercury is at long wavelengths.

Charged Particle Radiation--The radiation environment and shielded levels for the mission have been calculated for a range of shield thicknesses and two model geometries. Statistical boundaries were chosen to achieve a 95% confidence that these levels would not be exceeded. With 200 mils of aluminum, the levels are $\approx$ 47 and $\approx$ 100 krad(Si) for double slab and solid sphere shields, respectively. A 3-D shielding analysis has not been carried out for the actual spacecraft geometry, but it is likely that results from the double slab and spherical shield models would bracket the results of a 3-D analysis. The radiation design capability of spacecraft assemblies is required to be at least twice that of the 3-D shielded dose, providing a radiation design margin (RDM) of $\geq$ 2.

Reliability--Engineering subsystems and science instruments are required to meet at least Class "A" and "B" standards, respectively.

Spares--A set of single spares, tested through the assembly level, and selected long-lead-time components are required. If N units of a particular high-level assembly are needed to cover the total needs of one spacecraft, then 2N+1 units are required.

System-Level Conceptual Design

A novel system design that utilizes conventional technologies has been developed for the Mercury Orbiter and meets the spacecraft requirements while avoiding use of high-cost, high-risk, exotic technologies. Features of the design are briefly summarized below and then more detailed information is provided later.

Low-phase-angle, low-altitude passes over the Mercury surface expose the spacecraft to intense heating from the planet as well as from the Sun. In combination, the resultant flux covers a broad wavelength range and bathes virtually the entire exterior spacecraft surface. To circumvent problems associated with this most severe environment, the spacecraft is designed to insulate itself shortly before exposure and maintain that state until the flux has dropped to an acceptable level. Fortunately, the 12 hr elliptical orbit minimizes the exposure time, limiting the necessary "covered" period to a maximum of $\approx$ 3/4 hr. The insulation together with short exposure time and high heat capacity, i.e., high "thermal inertia" of the spacecraft prevent the internal temperature rise from being excessive. Use of a similar self-insulating capability prevents excessive temperature fall during occultation periods.

The design, similar to that employed in Helios, orients the spacecraft spin axis perpendicular to the Sun direction. This has multiple benefits: 1) it provides Sun-free top and bottom areas that are then used for radiative cooling, 2) it effectively distributes the incident solar flux around the spacecraft perimeter helping to isothermalize the spacecraft and minimizing peak temperatures, and 3) it reduces the range of necessary articulation of the high-gain antenna in one axis.

Single-axis solar arrays are employed that articulate in cone angle with respect to the spin axis. This allows control of solar input to the panels and, in turn, their electrical power output and temperature.

The design integrates the propulsion subsystem structure with that of the electronics bus, but continues to allow independent assembly and test of the electronics. This results in a compact design for the central body that reduces mass, lowers the height and center of gravity in the launch vehicle (allowing two spacecraft to be stacked), and provides an oblate structure with a constant center of gravity location that simplifies design of spacecraft dynamic control.

Structure/Configuration

The spacecraft flight configuration is shown in the frontispiece of Section IV, and Figure 13 provides internal views of the spacecraft central body. The primary support structure is a truss. This structure is augmented by 8 triangular top/bottom plates, portions of which also serve as radiators; 4 near-in sensor mounting plates, that also provide thermal interfaces for these sensors; and 4 side plates. Note, that internal

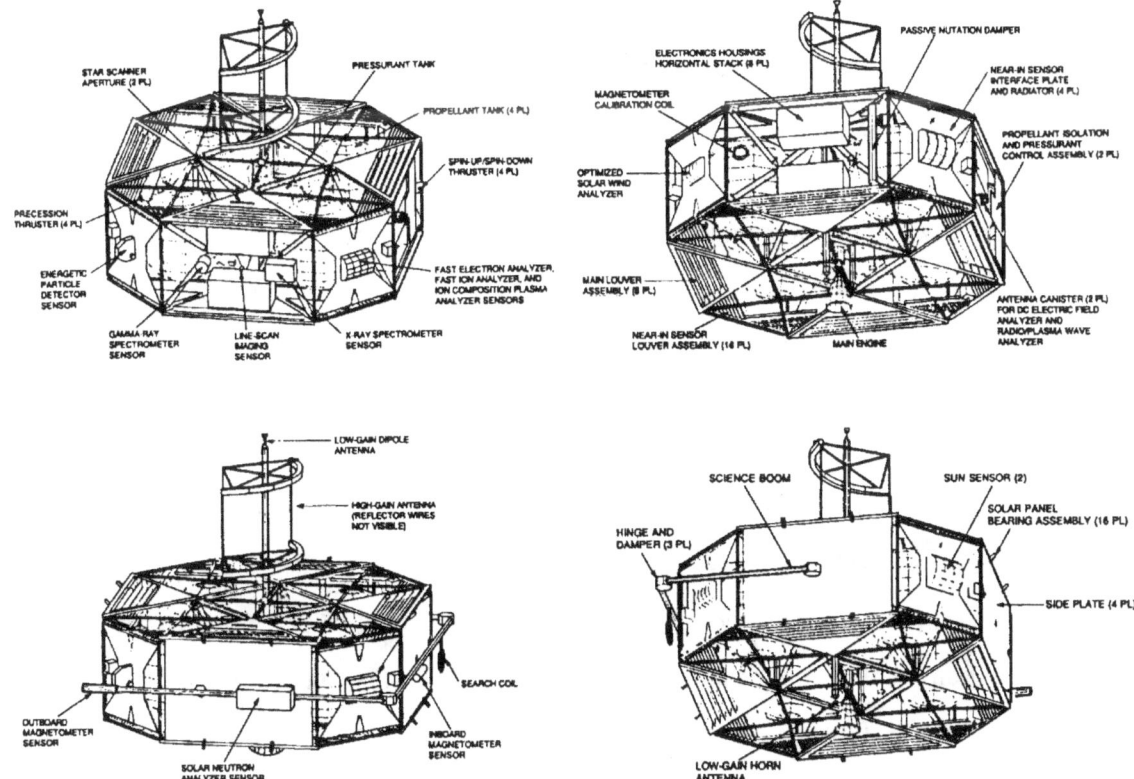

Figure 13. Spacecraft configuration with booms stowed (shown without solar panels, engine shade/radiator, and blankets). Lower views include side plates and science booms.

science sensors, the wire boom assemblies, and the science boom mount to the side plates. A tubular structure is used for the science boom and is wrapped partially around the spacecraft central body during launch. A conical structure at the base of the spacecraft is used for the main engine Sun shade and shunt radiator.

Science and engineering electronics are located in horizontal stacks of individual aluminum housings that are attached to the main thermal radiators and side plates; a horizontal strap attaches to each housing and provides additional lateral support. The perimeter of each housing is standardized at 28 cm on a side.

The width of each housing is varied as necessary to accommodate the needed electronics volume. Grounded foil interference shields separate adjacent housings, and square end plates on each housing stack provide axial electromagnetic and radiation shielding.

A 200 mil thickness for the perimeter walls and stack end plates is used and simultaneously provides adequate structural support, heat conduction, heat capacity, and radiation shielding. Note that the mass for instrument housings comes out of the total mass allocation for each instrument. During spacecraft assembly and testing operations each individual electronics housing is installed as available or as needed by placing it in its predetermined position in a particular stack, attaching it to the radiator plate, attaching it to the horizontal strap, and mating its connectors with test connectors or those in the cable harness. In later assembly stages it is also connected through standoff spacers to the spacecraft side plate.

Temperature Control

The spacecraft thermal design is primarily based on controlling how heat flux is distributed on the spacecraft surface, minimizing undesired heat input and transfer, controlling heat rejection, maximizing effective heat capacities, and isothermalizing interiors. These design features are briefly summarized below. Then, specific areas are discussed in more detail.

Spinning the spacecraft effectively distributes the solar heat flux and, in most instances, the planetary heat flux around the perimeter. This reduces the peak surface temperatures of sensors, solar cells, and the sides of the spacecraft and helps isothermalize the interior.

Low-solar-absorptance/high-emissivity, electrically-conductive exteriors are used to reduce temperatures of most surfaces that are exposed to direct or indirect solar flux and consequently reduce the solar heat input. In areas used for cooling on the top and bottom of the spacecraft, direct solar heat input is eliminated by keeping the Sun direction parallel to the radiator surfaces and recessing the radiator/louver assemblies. Heat flux from Mercury is prevented from excessively warming the radiators by covering them with insulation when the spacecraft is at low altitude over the hot surface.

High-efficiency, high-temperature, multilayer insulation, i.e., MLI or "thermal blankets", minimize undesired heat transfer. Both heat flow into the spacecraft during most of the Mercury orbit and heat loss in early cruise and in occultations at Mercury are minimized. Holes in the blankets for sensor apertures are as small as possible. Sun shutters further reduce heat input through the imaging and x-ray spectrometer apertures.

Electronic heat dissipation in the body-mounted and boom-mounted sensors is minimized by locating as much of their electronics as possible in housings in the spacecraft central body. This reduces peak heat loads on the sensors and makes it simpler to provide heat rejection and temperature control of the electronics.

Effective heat capacities of temperature-sensitive units are maximized by providing good conductive and radiative coupling to the spacecraft structure, housings, propellant, and other high-mass, high heat capacity elements in the spacecraft interior. Near exterior walls, however, the interior surfaces should have high emissivity surfaces to maximize internal radiative heat transfer. This tends to isothermalize the interior and allows an averaging over time that minimizes temperature extremes.

Central Body

Louver-covered radiators on the top and bottom of the spacecraft are used for heat rejection from the spacecraft central body. Since there would be a net heat input to these radiators when the spacecraft is at low altitude over the hot side of Mercury, the radiators and louvers are covered with insulation during these periods. This is accomplished by rotating the solar arrays. Each solar panel has solar cells on one side and multilayer insulation on the other. The panels can be articulated and are used for power generation when the normal vectors to their surfaces are at cone angles between 0° and +90° with respect to the spin axis on their side of the spacecraft. When it is necessary to cover the radiator/louver areas, the panels are rotated to -180° positions. This results in the cell side of the panels facing in towards the radiator/louvers and the

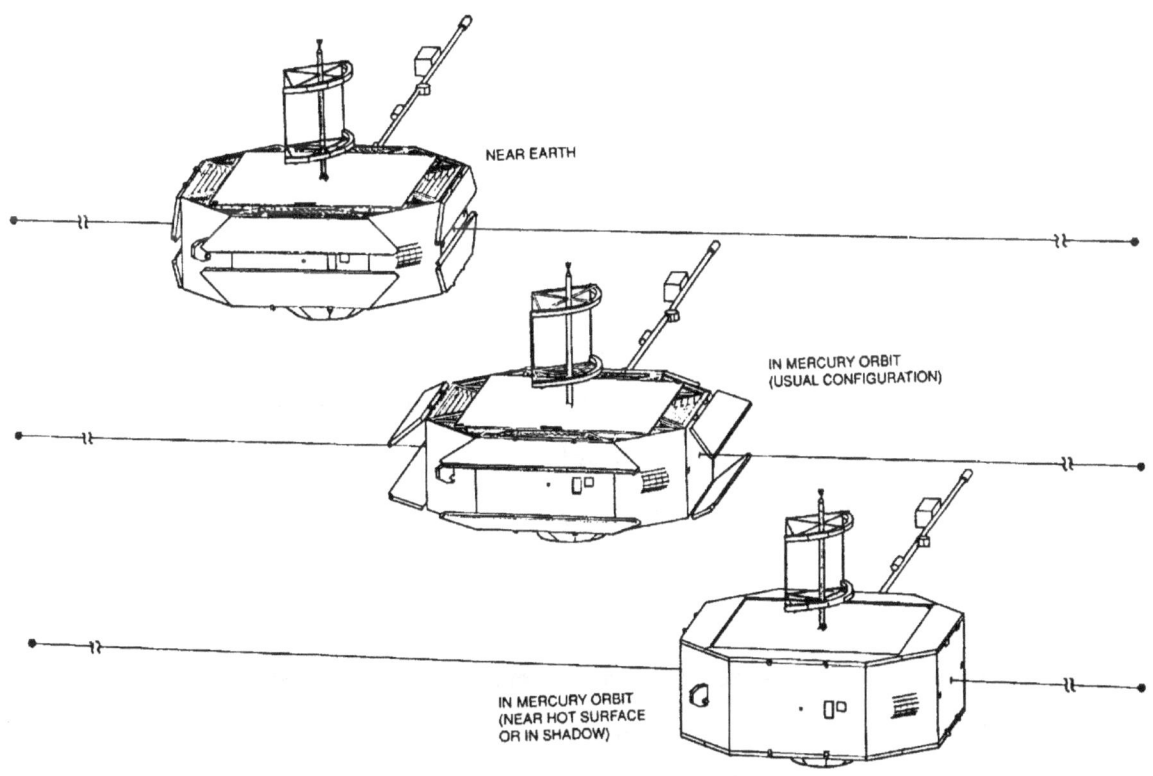

Figure 14. Spacecraft with solar panels shown in three positions.

insulation side facing outward. This has the combined benefit of protecting both the cells and the radiators from overheating. The solar panels can also be rotated to closed or partially closed positions when the spacecraft is in occultation to prevent extreme cooling of the solar panels and excessive heat loss from the central body. Figure 14 shows the range of positions of the solar panels in flight.

The horizontal stacks of electronics housings are attached to the main radiators. Heat from leakage and internal dissipation is rejected by the radiators. Together with louver assemblies that vary effective radiator emissivity, they keep internal spacecraft temperatures within acceptable limits.

Near-in Sensors

Additional louver-covered radiators on the top and bottom of the spacecraft are used for heat rejection from the near-in sensors. These, too, are covered by the solar panel blankets when the spacecraft is near the hot Mercury surface and the panels are in their closed positions. They also could be covered or partially covered when the spacecraft is in occultation.

Near-in sensors are mounted to separate thermal interface assemblies on each of the four sides of the spacecraft that are nearest the propellant tanks. The radiators and their louver assemblies are recessed from the top and bottom of the spacecraft to avoid solar heat flux. An electrical heater is attached on the other side of the interface plate from the sensor mounting surface and supplies a small amount of heat in early cruise and during occultation to prevent the sensor from getting too cold.

Boom-Mounted Sensors with Moderate Temperature Limits

The side wall area around the perimeter of each boom-mounted sensor, i.e., the area exposed to solar flux, is minimized relative to total surface area to reduce solar heat input. Small radiators located on the top and/or bottom surfaces are Sun free and are shaded from heat flux from the spacecraft central body. Inputs from internal and exterior transducers are used by the instrument electronics in determining when to close and open insulated covers over the radiators. Covers are opened under the combined conditions wherein the internal temperature is above a predetermined upper threshold (TU) and the net heat flow will be outward. Covers are closed under either of two conditions: 1) when the internal temperature is below a predetermined lower threshold (TL), and 2) when the net heat flow would be inward.

Nominally, as the spacecraft leaves the vicinity of the hot Mercury surface the sensor temperature is above TU and the covers are opened immediately when it is verified that the net heat flow would be outward. The covers are then closed when the sensor temperature has dropped to TL, which normally occurs before the spacecraft has returned to Mercury and reached the other heat flow reversal point. Under some conditions it may be possible for the spacecraft to reach this heat flow reversal point while the temperature is still above TL; this will also result in cover closure.

In early cruise, the radiator cover is closed to avoid excessive heat loss, and, if sensor heat dissipation is very small, a small amount of supplemental electrical heat is added to maintain adequate sensor temperature.

As in the spacecraft central body, temperature variations of sensitive elements in external sensors are further minimized by closely coupling them to high heat capacity hardware and decoupling them where possible from sources of heat leakage such as the blanket and the science boom.

Boom-Mounted Sensors with Wide Temperature Limits

Thermal designs for boom-mounted sensors with wide temperature limits are very similar to those for sensors with moderate limits, and active control, except that the radiators are permanently exposed. The radiator has a net outward heat flow for each of the planned ≥ 12 hr elliptic orbits and is sized to offset the net heat input from leakage and internal dissipation.

Propulsion

Monomethylhydrazine fuel and nitrogen tetroxide oxidizer are used for propulsion. Four 89 cm diameter tanks provide a total propellant capacity of 1600 kg, of which 1556 kg is usable. A single cylindrical, filament-wound tank is used for the pressurant. A main engine provides large ΔV burns while eight much smaller thrusters are used for precession, spin-up/spin-down, small axial ΔV, and small lateral ΔV. Note, precession and spin-up/spin-down thrusters are fired in couples to avoid undesired ΔV.

Power

Electrical power is provided by 8 single-axis-articulated gallium arsenide/germanium solar arrays which provide $\approx$ 303 W at Earth and, at an angle limiting the peak flux to 2.6 solar constants, $\approx$ 415 W at Mercury. Regulated 30 V dc is supplied to users through solid-state switches that limit turn-on transients, provide adjustable over-current protection, and allow monitoring of the state of each switch and its load. Three 8 Amp-hr nickel cadmium batteries provide energy storage for periods when the solar panels are closed near Mercury and for occultations. Continuous operation on batteries can extend 130 min.

Central Computer

A central computer provides processing for all the engineering subsystems. The specialized science processing is contained in the instruments. The computer uses the radiation-hardened SA3300 microprocessor family, includes direct memory access, and is capable of 300,000 instructions per second. The computer memory has 32 kbytes of PROM and 512 kbytes of RAM with built-in error detection and correction. Communication with science instruments is through a 1553B bus, and packetized telemetry is provided by the instruments. Two 1.8 Gb digital tape recorders provide data storage.

Attitude and Articulation Control

The launch vehicle spins both spacecraft up to a few RPM immediately prior to separation, and then after separation, the spin-up thrusters on each spacecraft are used to increase the rate to 20 RPM. This rate is then used during cruise and most of the orbits of Mercury. The spin rate for the planetology-intensive orbits is reduced to 10 RPM for improved compatibility with the planetology investigations. Spin axis orientation is always maintained at right angles to the Sun direction with the possible exception of short duration re-orientations to facilitate surface imaging (e.g., see Figure 8b).

Spacecraft attitude information is provided by Sun sensors and star scanners. The Sun sensors are mounted on the side of the spacecraft and provide Sun cone and clock angles with respect to the spin axis. The star scanners use charged-coupled device line arrays that view $\approx$ 0° to 20° from each of the two spin axis directions. Upward or downward fields of view are selected as necessary to avoid viewing Mercury and to allow the scanners to provide spacecraft clock angles with respect to the Sun line.

Control is maintained of the high-gain antenna despin motor speed and phase, high-gain antenna tilt angle, solar panel angle, Sun shutters, small thrusters, and main engine. A passive nutation damper is included to minimize spacecraft wobble.

Telecommunications

The command and telemetry subsystems use X-band transponders and 10.6 W solid-state power amplifiers. The high-gain antenna is normally used for commanding and always used for high-rate telemetry. It is based on the Helios design, but is scaled up in size and frequency. Low-gain horn and dipole antennas are provided for emergency commanding. The link capability at Mercury is 9 to 64 kbps for a 34 m deep space station, depending on range. A link with a 70 m DSS can support 40 to 276 kbps, but rates above $\approx$ 100 kbps are not available due to central computer constraints. A single DSS can cover both spacecraft as long as the angular distance between the spacecraft is not excessive. Expected 34 m Deep Space Network (DSN) coverage is one 8 hr pass every 3 days in cruise, and one 8 hr or two separated 4 hr passes per day at Mercury. The nominal data return strategy is likely to be based on using one 4 hr pass in each 12 hr orbit to return real-time data and data that was recorded throughout the rest of the orbit. The 70 m DSN is only expected to be available for critical engineering and science events.

Design Status

A viable spacecraft conceptual design has been developed and documented and satisfies the fundamental requirements on the spacecraft. Future pre-project studies should, among other areas, include: long wire boom dynamics impact analysis, high-gain antenna alternatives analysis, improved thermal materials identification, overall thermal design optimization, structure/adapter launch loads analysis, computer margins analysis, and cost estimation refinement.

4.3 MISSION DESIGN

Mission Synopsis

In designing a Mercury orbiter mission, it is difficult to accommodate the competing objectives of the magnetospheric, planetological, heliospheric, and solar researchers in a single design, especially given realistic cost constraints. For example, the basic orbit design and spacecraft configuration requirements for a magnetospheric mission are quite different from those of an imaging mission. For an imaging mission, a nadir pointed, 3-axis stabilized spacecraft in a low-altitude circular orbit is preferred. This contrasts with a magnetospheric mission, where fast-spinning spacecraft in eccentric orbits are necessary to accommodate in-situ fields and particles measurements requiring that all directions be viewed while traversing the major regions of the magnetosphere.

The MeO mission has as its primary objective the study of the magnetosphere with the objectives of the other science disciplines being important, but secondary. Fortunately, the thermal control required for a spinner in an elliptical orbit in Mercury's harsh solar and planetary environment is far simpler and less expensive than for a three-axis stabilized vehicle. A novel flight system based on conventional technologies has been designed to satisfy the MeO mission needs.

The mission begins with a launch of two identical spin-stabilized spacecraft using a Titan IV with a Solid Rocket Motor Upgrade (SRMU)/Centaur upper stage launch vehicle. The launch opportunities of August 1997, or July 1999, are the first two suitable for the flight system designed. It is expected that the same flight system may be applicable for three to four other mission opportunities available in the years 2000 to 2010. The transfer time to Mercury is from 3.5 to 5 years, depending on the launch year.

Following injection into the transfer orbit from Earth, one of the spacecraft (SC-1) is given a course change so as to arrive at Mercury 2 to 4 days later than the second spacecraft (SC-2). This phasing serves to create a spatial separation between the two spacecraft to accommodate the correlated cruise experiments. To arrive at Mercury with a sufficiently low V_∞, both spacecraft will be subjected to two Venus swingbys and two or three Mercury gravity assists, depending on the launch opportunity.

Upon arrival at Mercury, the two spacecraft are to spend about 4 Mercury years to carry out well coordinated experiments through various mission phases. The emphasis is on magnetospheric experiments for the first 2 years, during which two spacecraft are placed in complementary elliptic orbits, one equatorial and one polar, to allow for simultaneous two-point measurements of the magnetosphere. At the end of the magnetospheric exploration, surface mapping is conducted from two polar orbits with periapsis locations at the north pole and at the equator, respectively.

During the MeO mission, extensive coverage of the 0.31 to 0.72 AU region is possible, and should lead to measurements complementing those taken during the HELIOS mission. It is also possible to have correlative studies involving the two spacecraft.

The anticipated contributions to solar physics include opportunities to observe the solar neutron flux, as well as x- and gamma-ray measurements for an extended period at close proximity to the Sun. Plasma wave and radio science experiments are also an important part of the mission.

Heliocentric Trajectory Characteristics

The MeO mission is made possible with conventional propulsion systems by judicious use of Venus and Mercury gravity assists. These trajectories involve two Venus swingbys and two or three Mercury swingbys prior to the Mercury orbit capture. They are designated as the E-VVMM-M and E-VVMMM-M paths. As an example, the heliocentric trajectory for the 1997 launch year is shown in Figure 15, which is an E-VVMM-M trajectory.

The path between Earth and Venus is a Type IV (1.6 revs) and the path between Venus and Mercury is a Type VII (3.4 revs). The locations of Earth at various event times are shown in Figure 15 as well to illustrate the Earth-spacecraft communications geometry. The trip time is 4.9 years.

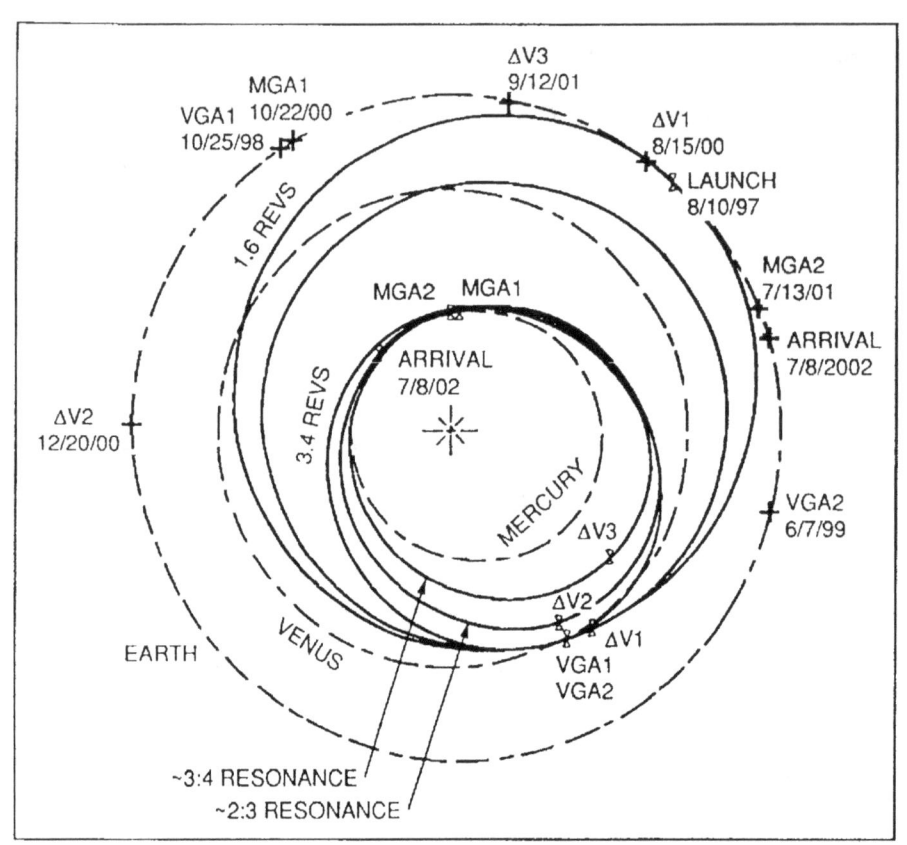

Figure 15. 1997 MeO trajectory (E-VV-MM-M)---ecliptic projection.

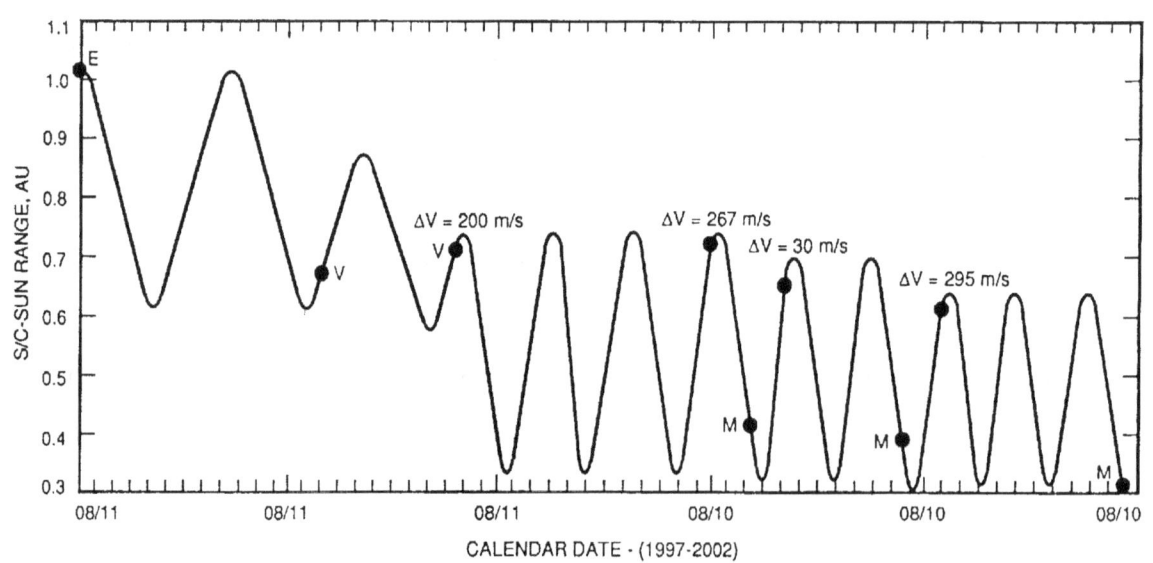

Figure 16. SC-2 solar range during cruise.

Figure 16 shows solar range as a function of time for one of the MeO spacecraft and presents a clear view of the effects of gravity assists on the trajectory.

In order to facilitate dual-spacecraft cruise experiments, a method of separating the two spacecraft must be devised. Since the mission is fashioned for a single launch, this spacing must come about at the expense of some ΔV maneuvers in deep-space. One way of controlling the spacecraft separation is by staggering the arrival times at Mercury. This has the added benefit of separating the coverage of two critical orbit insertion events in the flight operation. In this report, the spacecraft designated as SC-1 is controlled to arrive at Mercury on an off-optimal arrival date, while the other spacecraft, SC-2, is assumed to arrive on the optimal date as shown in Figure 15. If SC-1 is made to arrive 2 days off the optimal, an additional ΔV expense of about 70 m/s for the mission is incurred. It will be about 280 m/s for a 4-day separation. The relative spatial separation one can expect is of the order of a few million kilometers during the heliocentric transit.

Mercury Phase Mission Design

The scientific needs for this mission are met by employing several specific orbital stages during the mission. One spacecraft, SC-1 is captured into a relatively tight polar orbit, and permanently stationed there to survey the planet's magnetic field and to obtain solar wind data while the other spacecraft, SC-2, is exploring the far reaches of the magnetosphere and supporting planetological science objectives.

The orbit design strategy for SC-2 starts with a very loose elliptic orbit in a near-equatorial plane to allow for a survey of the far reaches of the magnetotail, out to more than 80 R_M, followed by reductions in the orbit size to cover the intermediate ranges. Once the spacecraft samples the far- and mid-ranges, its orbit is changed to a polar orbit to support near-range sampling and planetological studies.

As described in the next few subsections, this strategy results in the division of the mission phase into four subphases--Phase I: Orbit Insertion Phase, Phase II: Far-Tail Excursion Phase, Phase III: Mid-Tail Sweep Phase and Phase IV: Planetology Phase.

Phase I: Orbit Insertion Phase

For each spacecraft, the arrival V_∞ and aimpoint at Mercury determines the orientation of the respective initial capture orbit. The aimpoint for SC-1's polar orbit is near the north pole, while SC-2 is targeted for Mercury's dark side in the equatorial plane to set up for a loose equatorial elliptical orbit. Initially, after the orbit insertion, SC-2 drifts in regions of space uninteresting to magnetospheric investigations (see Figure 19) because the Sun is not properly positioned relative to the orbit orientation. However, in about a month, the Sun shifts to a position suitable for initiating a magnetotail exploration. It is envisioned that instrument calibrations, flight system health checks, orbit determinations, and orbital adjustments will be made during the month-long waiting period, in preparation for the high-activity phases to follow.

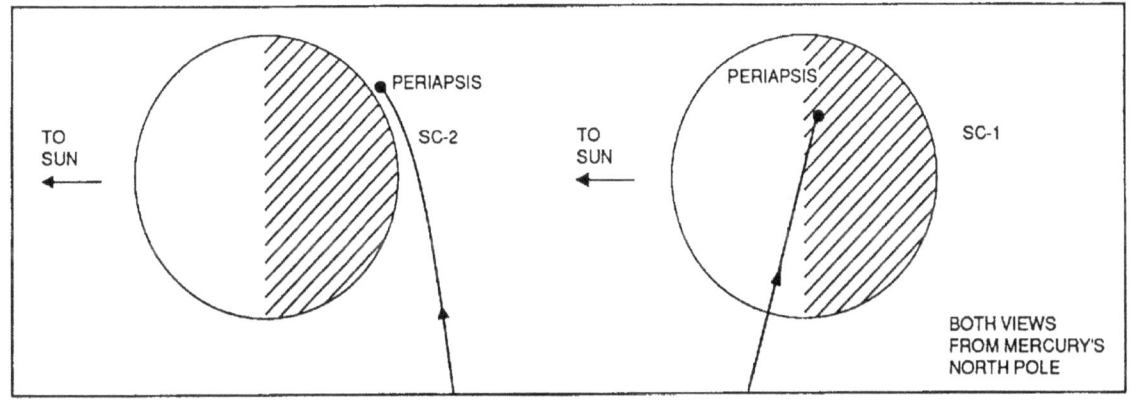

Figure 17. Mercury approach geometry of SC-1 and SC-2.

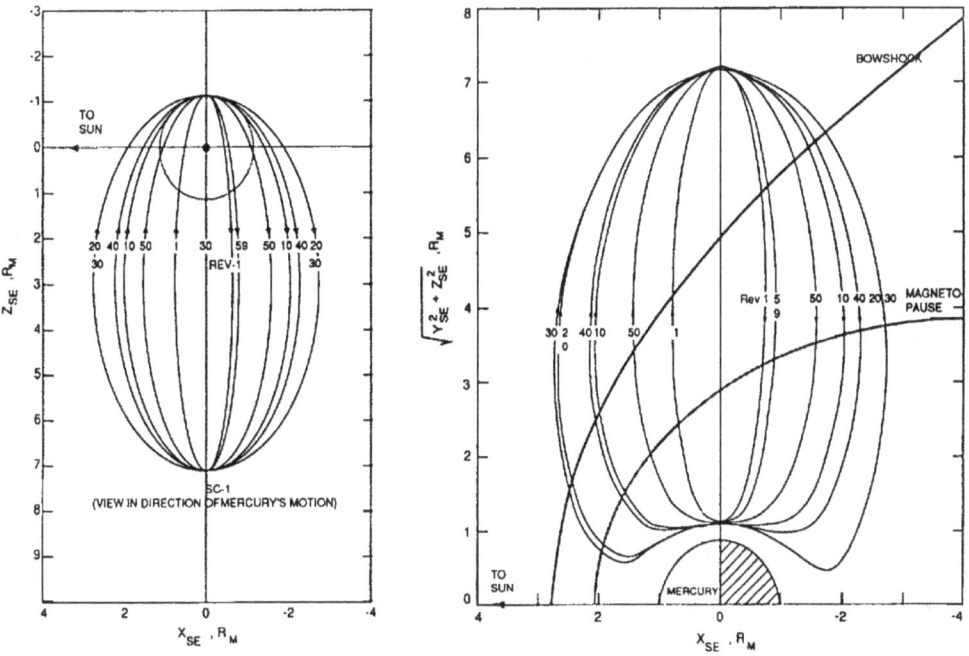

Figure 18a. SC-1 orbit. Figure 18b. SC-1 coverage of magnetosphere.

The approach conditions for SC-1 and SC-2 are shown in Figure 17. The captured orbits are shown in Figures 18a and 18b. Figures of orbits are given relative to the Sun in solar ecliptic coordinates to highlight the relationship of orbits to the magnetosphere. SC-1 is captured into a 200 km

X 6.2 R_M altitude, 12 hr polar orbit with the periapsis above Mercury's north pole. No significant changes to this orbit are planned for the remainder of SC-1's mission life. Figure 18b shows how this orbit cuts through the bowshock and magnetopause over time. Note that when the spacecraft orbit is not in an equatorial plane, the illustration of spacecraft magnetospheric coverage such as is seen in Figure 18b, is made using cylindrical coordinates.

SC-2 is initially injected into a 200 km X 83.7 R_M altitude, near-equatorial retrograde orbit, with a period of about 32 days. This injection condition is chosen to align the orbital geometry with the magnetospheric tail for as long as possible during subsequent phases of the mission. The shape of this orbit is substantially altered over time by solar gravity perturbation, and its inclination also changes by several degrees. SC-2 nominally will remain in its initial loose orbit for only a single cycle or about one month. Figure 19 shows the orbit of SC-2 in Phase I and continues into Phase II as seen from the north pole direction.

Phase II: Far-Tail Excursion Phase

As SC-2 returns to periapsis after one revolution in its initial loose orbit, the Sun-relative orientation of the orbit becomes suitable for the spacecraft to travel into the anti-Sun region of Mercury, as seen in Figure 19. This transition occurs at an altitude of about 15,000 km, and does not require a maneuver. In its second revolution around Mercury the orbit period is about 17 days. In this orbit, which is also nearly equatorial, SC-2 reaches far into the magnetotail out to over 80 R_M. It appears to linger in the 50 to 80 R_M region for 11 of the 16 days in this highly-

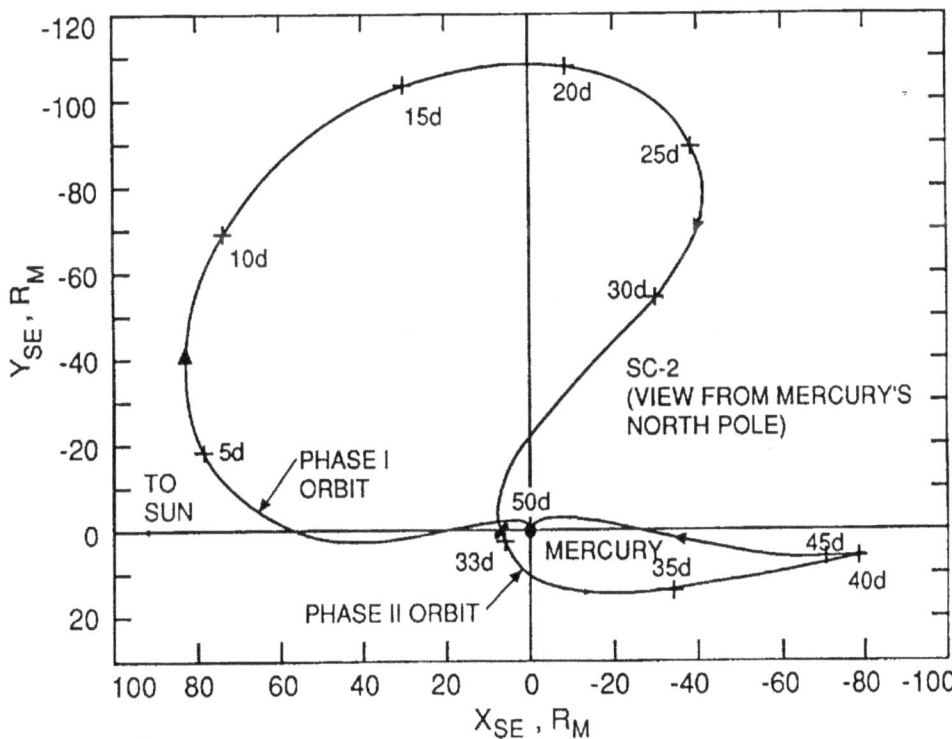

Figure 19. SC-2 orbit during Phase I and Phase II.

eccentric orbit while it is near apoapsis. At apoapsis, a maneuver is executed to adjust the upcoming periapsis altitude down to 200 km.

Phase III: Mid-Tail Sweeping Phase

At roughly 50 days after Mercury orbit injection, SC-2 completes its far-tail excursion phase and begins its mid-tail sweep. The transition is marked by a maneuver, executed near Mercury at periapsis, to slow SC-2's velocity and reduce its apoapsis altitude from over 80 R_M to about 32 R_M. The orbit continues to be near-equatorial, with a period reduced to 4 days. The spacecraft remains in this orbit for 24.5 revs--slightly over one Mercury year--to allow for two separate periods of multiple crossings of

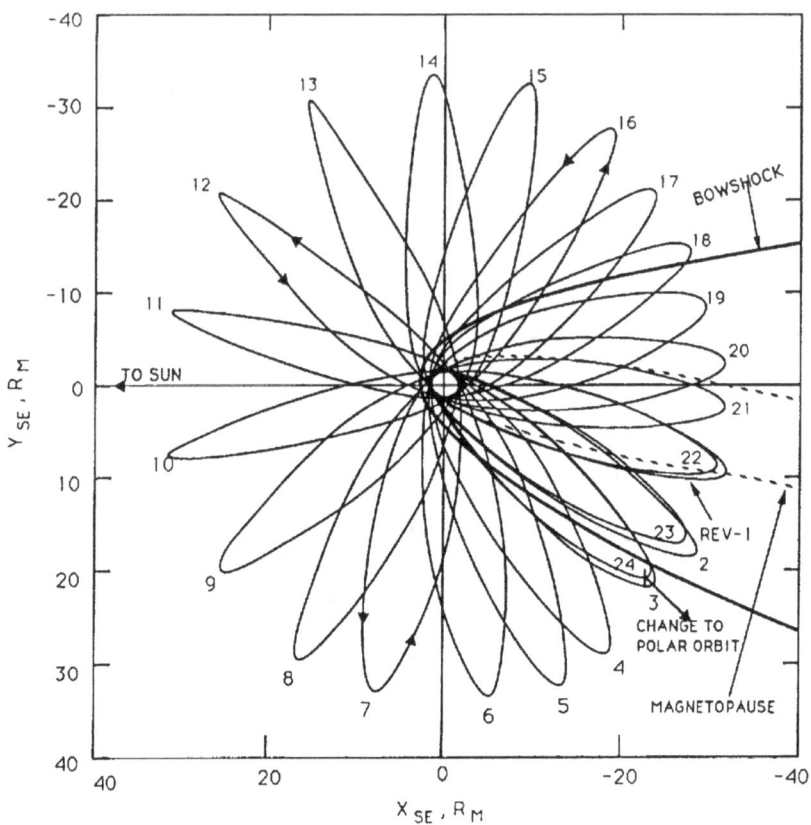

Figure 20. SC-2 orbits during mid-tail sweeping phase.

the ~30 R_M magnetotail region. The spacecraft appears to orbit Mercury in a flower-petal pattern when viewed in Mercury-Sun fixed coordinates as shown in Figure 20.

Since the orbit is nearly equatorial, SC-2 goes through a period of solar occultation in each orbit. The time SC-2 spends in Mercury's solar occultation zone starts to dramatically increase at about rev 15 and, if unadjusted, could grow to a 10 hour occultation at rev 20 when the apoapsis is in the anti-Sun direction. Spacecraft energy needs for this long a period without solar power are beyond the storage capabilities of reasonable, low mass batteries. Therefore, to avoid these intolerably long occultations, a maneuver is executed one day before the start of rev 18. This adjusts the orbit nodal line to being nearly perpendicular to the Sun direction and also changes the inclination from $\approx 0°$ to 6°. The action costs 90 m/s of ΔV but it manages to shift the position of SC-2 just outside of the umbra but still inside the nominal tail boundaries when SC-2 is near apoapsis. This reduces the occultation period to fit within the spacecraft design specifications of less than 130 min.

Halfway through rev 25 (at apoapsis), SC-2 executes a 194 m/s ΔV maneuver to flip the orbit plane to a polar orientation while retaining a 4 day period. This sets up SC-2 for the final phase of its mission.

Phase IV: Planetology Phase

As SC-2 rounds periapsis on rev 25 of the previous phase, a ΔV maneuver of 211 m/s is made to reduce the orbit period from 4 days to 12 hr. Note that the final orbit period can be any value greater than 12 hr, if adjustment of relative phasing between the two spacecraft is desired. The location of the periapsis of the resultant orbit is at the equator allowing SC-2 to complement SC-1 in planetology. Both spacecraft are expected to remain in these polar orbits for two full Mercury years, 176 days, during which Mercury rotates three times under these orbits.

Relative orientations of SC-1 and SC-2 are illustrated in Figures 21a and 21b. Figure 21a shows the first few revs of orbits viewed in the direction of Mercury motion. Figure 21b shows that the orbital planes of the two are separated by about 65°. The differences in periapsis locations

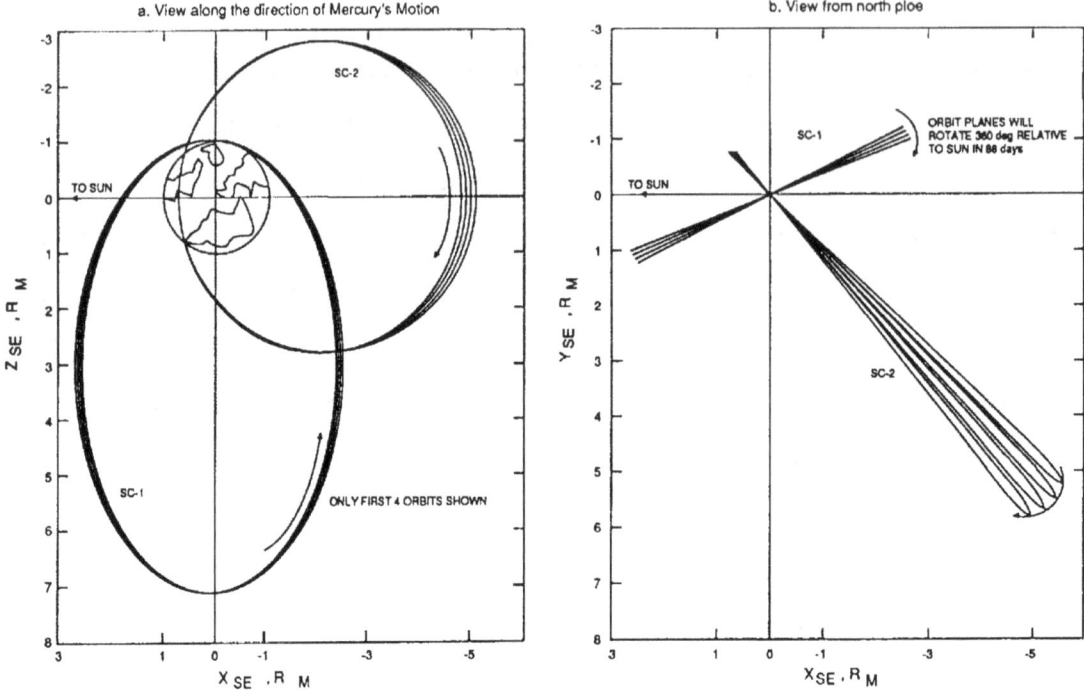

Figure 21. SC-1 and SC-2 orbits during planetology phase.

and the separation of the orbit planes are advantageous for the planetology objectives.

Both spacecraft will carry CCD cameras. Imaging is to be made with the spin axis of each spacecraft pointed parallel to Mercury's spin axis; the assumed spin rate for this analysis is 10 RPM. The CCD array itself consists of 2048 x 1 pixels with 13-micron pixel size. A 50 mm focal length, f/4 optical system will be used to focus images on the array. This system provides an instantaneous field of view of 0.26 mrad and a half-angle field of view of 14.9 degrees.

The mission requirement is to image 25% of the surface at 100 m or better and to obtain full coverage at better than 1000 m. Mercury rotates once in 58.6 days, so adjacent ground tracks will only be separated by 3° of longitude. Thus, consecutive orbits will provide longitudinal overlap without severe foreshortening.

For each spacecraft, the camera will point to nadir only at the equator. At other latitudes, non-nadir pointing degrades the imaging resolution. SC-1 will provide excellent coverage of the northern hemisphere; since its periapsis is over the north pole, the high latitudes will be imaged from low altitude. This spacecraft provides 1 km or better resolution (i.e., 2 pixels) coverage of Mercury from 8°S to the north pole. The north pole itself will be imaged at 640 m resolution. For SC-2, the best resolution (104 m) is obtained at the equator, where the spacecraft is at periapsis and the camera is nadir-pointing. The resolution decreases from that point, and drops to 1 km resolution at 77°S and 77°N.

Overall, more than 60% of the planet will be imaged with resolutions better than 250 m, and 98% at 1 km or better. Neither spacecraft can image the south pole area at 1 km resolution from its nominal orientation. However, better coverage of this region is possible with occasional spacecraft attitude changes which improve the look angles to the surface. For example, when the spacecraft is in a near-terminator orbit, the thermal environment is such that the spin axis can be tilted to expose radiator areas of the spacecraft to Mercury without the loss of thermal control. In principle, any attitude change is permissible so long as the spin axis is maintained perpendicular to the Sun and one avoids exposing the top and bottom of the spacecraft to the hot Mercury surface. However, concerns for the loss of communications during the off-nominal orientations and for excessive fuel demands for attitude changes will limit the number of such maneuvers.

Since the orbit plane of a polar orbit is fixed in inertial space and because of the 3:2 spin-orbit coupling of Mercury, each area of the Mercury surface will be imaged three times at three differing phase angles by each spacecraft during the two Mercury-year imaging phase. Continued imaging beyond two Mercury-years will not provide different phase angles unless the orbit geometry is changed.

Mission Performance Requirements and MeO Spacecraft Capability

Propulsive requirements for the mission scenario described above and the capabilities of the MeO spacecraft are summarized in Table 7 for the 1997 and 1999 launch opportunities. A conservative launch period of 20 days is assumed here although a 10-day launch period is generally considered adequate when using an expendable launch vehicle.

The capability of the current Titan IV (SRMU)/Centaur launch vehicle and the propulsion capability of the proposed system design enables the performance of this mission with comfortable launch margins and adequate propellant or payload margins. As detailed in Table 8, the end-of-mission mass for the proposed spacecraft design is 800 kg.

Mission Opportunities for 1997-2010

Currently known mission opportunities available from 1997 to 2010 are given in Table 8. The performance possible with the proposed spacecraft is shown for the best launch date of each opportunity. Those missions inside

Table 7. MeO Mission Requirements and Performance Summary

	1997 LAUNCH		1999 LAUNCH	
C3 REQUIREMENT (FOR 20-day LAUNCH PERIOD)	14.15 TO 15.64 $(km/s)^2$		17.37 TO 22.2 $(km/s)^2$	
ΔV REQUIREMENTS (km/s)	SC-1	SC-2	SC-1	SC-2
HELIOCENTRIC				
DETERMINISTIC	0.882	0.844	1.364	1.314
NAVIGATION	0.200	0.200	0.225	0.225
MERCURY PHASE				
DETERMINISTIC	1.640	1.784	0.971	1.113
NAVIGATION	0.250	0.250	0.250	0.250
TOTALS	2.972	3.078	2.810	2.902

	1997 LAUNCH	1999 LAUNCH
LV INJECTION CAPABILITY	7334 kg	6583 kg
MARGIN FOR ERROR	-750	-750
	6584	5833
PROPELLANT MASS	1600	1600
MAX USABLE PROPELLANT	(1556)	(1556)
MAX EOM MASS*	835	920
TOTAL FOR TWO S/C	4870	5040
LV ADAPTER MASS	200	200
TOTAL INJEC REQUIREMENT	5070	5240
LV MARGIN	1514 kg	593 kg
EOM MASS MARGIN	35 kg	120 kg
OR ΔV MARGIN	78 m/s	255 m/s

* MAXIMUM THE 1556 kg OF PROPELLANT CAN DELIVER FOR THE GIVEN ΔV REQUIREMENT

Table 8. Launch Opportunities and Performances for Years 1997-2010

TRAJECTORY TYPE: E-VVMM-M						
LAUNCH YEAR	1997	1999	2002	2004	2005	2007
LAUNCH DATE	8/10	7/25	9/07	7/10	8/05	7/09
C3 $(km/s)^2$	14.1	18.1	11.2	28.4	16.6	21.2
TOTAL ΔV (km/s)	3.08	3.27	3.66	3.02	3.00	3.35
FLIGHT TIME (yr)	4.9	3.4	3.0	3.8	4.2	3.7
EOM MASS (kg)	860	753	614	861	872	723
EOM MASS MARGIN (kg)	60	-47	-186	61	72	-77
LV MARGIN (kg)	1593	1342	2439	52	1277	1062
TRAJECTORY TYPE: E-VVMMM-M						
LAUNCH YEAR	1997	1999	2002	2004	2005	2007
LAUNCH DATE	8/10	7/25	9/07	7/10	8/05	7/09
C3 $(km/s)^2$	14.1	18.1	11.2	28.4	16.6	21.2
TOTAL ΔV (km/s)	2.58	2.88	3.21	2.64	2.60	2.89
FLIGHT TIME (yr)	6.4	4.9	4.5	5.3	5.6	5.2
EOM MASS (kg)	1108	933	778	1036	1098	927
EOM MASS MARGIN (kg)	308	133	-22	236	298	127
LV MARGIN (kg)	1097	982	2113	0	823	654

• COMPARISON BASED ON 1-day LAUNCH PERIOD

the boxes are the viable ones for the proposed spacecraft design. It is anticipated that with a revised trajectory optimization method, and the use of a post-launch ΔV optimization instead of a total ΔV optimization, it will be possible to utilize the 2002 opportunity. Otherwise a slight modification of either the mission requirements or the spacecraft design may be required.

V. FUNDAMENTAL PROBLEMS IN PLANETARY SCIENCE ADDRESSED BY MeO

5.1 MAGNETOSPHERIC PHYSICS SUMMARY

Solar Wind-Magnetospheric Coupling

The solar wind populates and energizes the magnetospheres of Earth and Mercury. This contrasts with the magnetospheres of Jupiter, Saturn, and Uranus where internal sources dominate. Plasma processes operating in the boundary region between the magnetosphere and the solar wind execute these functions. At Earth, the boundary admits a few percent of the incident solar wind particles and energy. The way it admits them is still uncertain. It behaves as if it contained a variable portal through which the solar wind enters. How that portal opens and closes, what it looks like (there might be more than one), and even where it is are uncertain. Models address these issues, but none is comprehensive nor definitive. Data from Mercury can reveal entry processes and resolve entry structures in at least three ways: 1) provide synoptic observations of boundary structure, 2) resolve the origin of the plasma mantle, and 3) reveal boundary layer dependences on fixed boundary conditions.

A Mercury orbiter can skim the sunward boundary of this miniature magnetosphere from top to bottom or side to side before sufficient time passes for the solar wind to change significantly. Such passes will show the synoptic flow and the magnetic pattern across the sunward hemisphere. The distribution of accelerated flows and flux transfer event signatures can locate the site of magnetic merging unambiguously. This type of direct synoptic analysis is impossible at Earth because during the time a boundary-skimming satellite surveys the territory, the solar wind changes and the site of merging shifts. In the same way, the Mercury data will show how the different boundary layers fit together or merge into each other. At Earth the global configuration changes too fast. In going from one regime to another, a satellite most likely records a temporal event rather than a spatial gradient. Of course, the statistical or average positions of the different boundary layers are known for Earth, but the systematic shifting of their positions that must attend the configurational changes in the magnetosphere induced by the solar wind are unknown.

The plasma mantles, also known as the high-latitude boundary layers, are widely regarded as the regions that generate most of the voltage that powers magnetospheric activity. Despite their importance, their origin within the terrestrial magnetosphere is ambiguous. One model populates them with particles that enter the magnetosphere through the weak-field cusps on the sunward boundary. From there the particles bounce off magnetic mirrors before reaching the Earth and shoot back up into the magnetosphere on field lines that are carried tailward by the general magnetospheric convection. The locus of their windswept ricochets is a fan which rubs against the magnetosphere's polar boundaries and defines the plasma mantle for each hemisphere. In another model, plasma mantles result from the solar wind streaming obliquely into the vacuum of the magnetotail through magnetically "open" high-latitude windows that attach in some unknown way to the sunward merging region and extend from there far down the tail. If the first model is correct, Mercury's magnetosphere has no plasma mantle, because the planet fills most of its magnetosphere and buries the requisite magnetic mirrors deep inside itself. Mercury's surface intercepts the mantles' source. But if the second model is correct, Mercury's magnetosphere has a plasma mantle, and boundary-skimming orbits can reveal how they attach to the sunward merging region. A major ambiguity in magnetospheric physics would be resolved simply with data from Mercury.

Magnetospheric physicists could exploit the differences between the physical properties of the media at the outer and inner boundaries of these two terrestrial magnetospheres to test and guide theories of boundary structure. To illustrate the general principle by specific examples, consider a cross section through a magnetically closed magnetospheric boundary, which is usually composed of two lamina: an outer current-carrying layer identified by a sudden change in the magnetic field (the magnetopause) and an inner layer identified by distinctive plasma properties (the low-latitude boundary layer). The magnetopause is important because it separates the outside world and the inside world of the magnetosphere. It is also the initial condition in the time-dependent merging scenarios. Despite its importance, the physics behind its structure is still unknown. The low-latitude boundary layer is important because it is the home of one of the main current generators powering magnetospheric activity. Despite its importance, the best theory of its structure is basically untestable with data from Earth's magnetosphere alone.

The problem with understanding the magnetopause is that its thickness is much greater than the natural scale length given by the ion gyroradius. This means that the magnetopause is more than just a turn-around layer for solar wind ions. It has a charged particle population of its own. Data on the thickness of a magnetopause formed in solar wind conditions distinctly different from those at Earth (e.g., 0.3 vs. 1.0 AU) would be immensely valuable to get a bearing on what determines the intrinsic magnetopause population. At Mercury, the average solar wind density is an order of magnitude greater than at Earth. The average field strength at the boundary is a half an order of magnitude greater. The many traversals of the magnetopause that will accumulate because of the rapid magnetosphere sample time at Mercury will permit a detailed analysis of the differences between the Hermean magnetopause and the terrestrial magnetopause as revealed by the ISEE and CLUSTER missions.

The most powerful theory of the low-latitude boundary layer ascribes its structure to an interaction between the boundary layer and the ionosphere. The natural scale length is given by a combination of ionospheric conductivity, the kinematic viscosity of the boundary layer plasma, and the amount by which field lines spread apart between the ionosphere and the boundary. The theory is nearly impossible to test at Earth because field-line spreading is large and its value is very uncertain. Further, the kinematic viscosity is an empirical parameter. The amount of spreading and the viscosity can be adjusted to give agreement between theory and observations. But this is no test. On the other hand, because Mercury fills most of its magnetosphere, field-line spreading is small as is the uncertainty in its value. The kinematic viscosity should be similar to that at the Earth. The ionospheric conductivity is at least an order of magnitude less. Thus, Mercury is a better place to test the theory, and the comparison with Earth strengthens the test.

Data from Mercury can also provide unique information on the general solar wind-magnetosphere interaction problem in at least two ways: 1) resolve an ambiguity in the cause of the sunward erosion of magnetospheres, and 2) document the properties of subsonic solar wind-magnetosphere interactions. Regarding the first, when the interplanetary magnetic field (IMF) points in a direction that favors merging with the planetary magnetic field in the sunward hemisphere, the sunward boundary contracts--a process called "erosion." Erosion can occur because the IMF penetrates the boundary and weakens the interior field or because the field generated by a merging-induced current running between the boundary and the ionosphere weakens the field near the sunward boundary. Magnetospheric models are not advanced enough to determine which mechanism dominates at Earth, though there is a bias for the second. But the scant data returned by Mariner-10 suggest that Mercury's boundary erodes relatively as much as the Earth's does. If so, the first mechanism must dominate, since strong currents linking the ionosphere can be excluded. A Mercury orbiter mission can be expected to settle this issue.

Although a possible subsonic solar wind interaction with Earth's magnetosphere was reported once, the probability of recording a subsonic solar wind interaction at Mercury is much greater. There is keen interest in studying such an interaction because with no bow shock there is no wave drag. This means that the only cause of upstream-downstream asymmetry in the interaction is tangential stress on the boundary. By eliminating a main cause of upstream-downstream asymmetry, a subsonic interaction reveals the strength of the tangential boundary stresses directly through the extent of the remaining asymmetry. This is an important measurement because tangential stresses extract energy from the solar wind to power internal magnetospheric processes.

Substorm Processes

Substorms are recurring global dynamical events that feed off magnetospheric convection. They are the main recurring dynamical events in Earth's magnetosphere. Despite their importance, their cause is unknown and there is not even an undisputed empirical model for the phenomenon. The problem is that the substorm engages too large a volume for too short a time for its life cycle to be documented by any practical constellation of spacecraft. Much has been learned by analyzing data taken by different spacecraft during the same events and through statistical analyses of data taken during many events. But the magnitude of the problem is dramatized by the fact that there are at least eight currently advocated theoretical models of the substorm.

Mariner-10 observations at Mercury revealed the presence of events with all the features of substorms that, if appropriately scaled in time and distance, would occur at Earth in the same instrument complement. A prime objective of a Mercury orbiter mission is to determine if the Mercury events are indeed analogs of Earth substorms. However, in framing specific scientific objectives for the mission, the evidence from Mariner-10 permits the assumption that they are substorm analogs. (If they are not, they become Mercury-specific or small magnetosphere-specific and take on importance in their own right.)

Substorm data from Mercury can eliminate substorm models more harshly than data from the magnetosphere that nurtured them. Earth substorms engage the ionosphere, and at least two strong contenders in the model competition put the ionospheric interaction at the heart of the process; i.e., no ionosphere, no substorm. Data from Mercury will provide stringent tests for these models. More than this, they will document substorm phenomena in a magnetosphere with a poorly conducting lower boundary. Any complete substorm model must be able to account for the phenomena in both high-conductivity and low-conductivity situations.

Because the natural time scale for the Earth substorm is about one hour--the same as the average time between significant solar wind changes--much confusion exists over the relative importance of internal and external triggers for the substorm. The substorm time scale at Mercury is a few minutes. Thus, data from Mercury should decisively resolve the question of internal or external trigger. (There could be both. If so, they will be well separated in the Mercury data.)

Because Mercury has a high substorm repetition rate, data taken with one satellite in the magnetosphere and the other in the tail would soon accumulate many events to test the popular-but-controversial hypothesis that the tail sheds plasmoids during substorms. (During substorm outbreaks, substorms recur roughly every 5 minutes--about the time a plasmoid takes to reach a satellite stationed at L2.)

At the planet's end of the tail, substorms inject energetic particles into the magnetosphere, whereupon as coherent packets, they circle the planet in less than 10 seconds and repeat until losses through the boundary deplete them. Here again the fast cycle time compared to the time for appreciable magnetospheric changes confers the advantage of relatively clean interpretations of the data. The injection process, the cycling process, the radial transport process, and the loss process can be examined under conditions approximating laboratory control. We can reasonably expect understanding of these basic processes gained at Mercury to result in greater understanding of the same processes at Earth.

Magnetospheric Calms

Whereas magnetospheric substorms occur when the IMF points in a direction that favors merging with the planet's magnetic field on the sunward boundary, magnetospheric calms occur when the IMF points in a direction to inhibit merging there. At Earth, magnetospheric calms produce not the absence of activity, but activity of a distinctive kind characterized by polar cap auroras and a curiously shaped polar cap. The magnetospheric structural counterparts to these known ionospheric manifestations of magnetospheric quiet are unmapped. The problem is that the condition of magnetospheric quiet is comparatively rare, because there is a lag of several hours between the time when the IMF acquires the requisite direction and the ensuing magnetospheric quiet. Usually, the IMF changes before magnetospheric quiet prevails. The terrestrial data needed to reconstruct the quiet plasma sheet configuration statistically take years to accumulate. Unlike the situation at Earth, Mercury's magnetosphere adjusts quickly to changes in IMF direction compared to the time between IMF direction changes. The plasma sheet geometry of a quiet magnetosphere will be seen as often as the geometry of the disturbed or post-disturbed plasma sheet; the normal states at Earth.

Magnetosphere-Ionosphere Coupling

The conductance of the ionized portion of Mercury's exosphere is estimated to be one to two orders of magnitude less than Earth's. On the other hand, the plasma content of a unit magnetic flux tube is comparable to that at Earth. The ratio of these quantities sets the strength of magnetosphere-ionosphere coupling. The comparison suggests that magnetosphere-ionosphere coupling at Mercury is stronger than at Earth. Consequently, at Mercury the convection electric field should be more strongly shielded from the inner magnetosphere. Coupling electric currents should flow (field-aligned currents), but they must be weaker because the driving voltage is less and the electrical conductance is smaller. These theoretical predictions must be made more quantitative. But they illustrate predictions must be made more quantitative, but they illustrate how data from Mercury can test basic magnetospheric theories that are nearly impossible to test at Earth because their key parameters do not vary enough. When there is only one case on which to test a theory and that theory works, the agreement could be accidental. But agreement in two very different cases is more convincing. And there is the even more exciting possibility of a disagreement.

Currents linking the magnetosphere and ionosphere at Earth play crucial roles in magnetospheric dynamics. To appreciate how important it is to get data from another terrestrial system with significantly different coupling parameters, we review the situation for Earth. According to the currently most comprehensive picture, the main energy conduit for the magnetosphere starts in the solar wind adjacent to the tail, passes through one or more magnetospheric boundary layers, courses through the main bodies of the tail and the magnetosphere and enters the ionosphere. Through the ionosphere it connects to a secondary conduit that exchanges energy with the particle populations of the inner magnetosphere, such as the ring current. From there, it takes a mirror-symmetric path back to the solar wind. In the ionosphere it also contacts electrically, and thus exchanges energy with the weaker cusp current system, which links the ionosphere to the part of the boundary that bears the main force of the solar wind. A fourth, and strong-but-sporadic current system (the substorm current wedge) links the mid-tail plasma sheet to the ionosphere and through the ionosphere to the other three current systems. The ionosphere is thus seen to be a contact center for routing information and energy between: 1) the main magnetospheric trunk line, 2) a branch line to the particle-filled inner magnetosphere, 3) an offshoot to the wind-pressed boundary, and 4) a sporadically engaged feeder line to the plasma sheet storm center. Thus, by their connections to the ionosphere, the main components of the magnetosphere communicate their actions and reactions to each other. Mercury represents an opportunity to apply harsh, quantitative tests to models of magnetosphere-ionosphere coupling--a central paradigm in magnetospheric physics.

Solar Wind-Planetary Coupling

Because Mercury fills most of its magnetosphere and because it has a large conducting core, sudden changes in solar wind pressure induce shielding currents on the core's boundary that generate a magnetic field strong enough to be detected by a Mercury orbiter. The induced field stiffens the magnetosphere's elastic response to sudden changes in solar wind pressure, which gives another way to measure the induced field. Measurements of the induced field can be inverted to infer the conductivity profile of the planet's interior, as was done for the Moon.

Occasionally, the solar wind at Mercury must blow hard enough to push the magnetosphere's sunward boundary down to the surface of the planet. In such instances, the solar wind contacts the surface in the subsolar region--the point where the wind's ram pressure pushes most directly against the boundary. But somewhere between the subsolar point and the terminator, where the ram pressure pushes most obliquely, the planet's magnetic field must re-emerge from the surface and project out into the wind. Since some of the solar wind is absorbed by the surface, less must be diverted around the planet. From the viewpoint of the solar wind, the planet appears to be smaller. Consequently, the bow shock moves even closer to the planet than if the wind merely compressed its magnetic field. Conceivably, the bow shock could, on occasion, become attached to the planet. The phenomena of such a magnetosphere are unknown.

In summary, it is seen that not only is Mercury highly valuable as a testing ground for magnetospheric understanding and as a new (and in some important ways a better) source of information on the structure and dynamics of terrestrial magnetospheres, but it also presents magnetospheric configurations that are unique in the Solar System. In all of these respects, Mercury Orbiter is the logical follow-on to the international missions which will conduct detailed investigations of the Earth's magnetosphere in the mid-1990s.

5.2 PLANETARY PHYSICS SUMMARY

Formation Process

How Mercury formed and acquired such a large iron core is a major unsolved problem in planetary science. Chemical equilibrium condensation models for Mercury's position in the solar nebula can account for an iron content resulting in an uncompressed density of only about 4 to 4.2 g/cm^3, rather than the observed 5.3 g/cm^3. Some other mechanism must have operated to concentrate this excess iron in Mercury. Until we understand this mechanism our knowledge of the origin of the terrestrial planets will be incomplete.

At present there is no way of deciding between the Selective Accretion, Post-Accretion Vaporization, or Giant Impact models invoked to explain Mercury's high mean density (see Section 2.2). However, each model predicts a significantly different composition for the silicate portion of Mercury. In the Selective Accretion Model, the differential response of iron and silicates to impact fragmentation and aerodynamic sorting leads to iron enrichment due to the higher gas density and shorter dynamical time scales in the innermost part of the solar nebula. The compositional consequences for this model are a refractory oxide abundance of between about 7 to 9 weight percent (alumina about 3.6 to 4.5 percent), alkali oxides about 1 weight percent, and FeO between 0.5 and 6 weight percent. In the Post-Accretion Vaporization Model an intense bombardment by solar electromagnetic and corpuscular radiation in the earliest phases of the Sun's evolution vaporized and drove off much of the silicate fraction of Mercury leaving the core intact. The compositional result is a severe depletion of alkali oxides and FeO (<0.1 weight percent) and extreme enrichment of refractory oxides (about 40 weight percent). A variation of this hypothesis is a Giant Impact of a planet-sized object that ejects much of Mercury's silicate mantle. Only about 10 percent of the ejected silicates are reaccreted, while most of the material is perturbed into orbits which eject it from the Solar System. The chemical consequences for Mercury's remaining silicate fraction are about 0.1 to 1 weight percent refractory oxides, between 0.01 and 0.1 percent alkali oxides, and 0.5 to 6 percent FeO.

The geochemistry investigation (XRF and GRS instruments) could provide the data required to decide between these competing hypotheses. Estimates of the global concentrations of the key elements Al, Mg, Si, Fe, Ti, K, Na and Ca can be used to infer the refractory and alkali oxide, and FeO content of Mercury's silicates. At the very least, these data will place constraints on any hypothesis for the formation of Mercury's large iron core, and consequently, the origin and early evolution of all the terrestrial planets.

Thermal History, Internal Structure and Crustal Dynamics

The thermal history of Mercury depends on the initial starting conditions. In one model, the planet forms in a molten condition and then cools over time. In the other model, Mercury forms homogeneously at moderate temperatures, heats up by the decay of radioactive elements and the migration of iron towards the center to form the large iron core, and then cools to the present time. The migration of iron toward the center is alone capable of raising the internal temperature by 700°C.

The presence of a dipole magnetic field strongly suggests that the outer core is currently in a fluid state. This requires a light alloying element in the core to lower the melting point, otherwise the entire core would have solidified long ago. The most likely candidate is sulfur. For a sulfur abundance less than 0.2 percent the entire core should be solidified at present, while an abundance of 7 percent results in an early fluid core formation. Therefore, the amount of core sulfur must be between 0.2 and 7 percent. Initial chemical equilibrium condensation models for Mercury's position in the solar nebula predicted the complete absence of sulfur, which is inconsistent with the presence of a partially molten core. However, three-dimensional computer simulations of terrestrial planet formation indicate that considerable mixing from the feeding zones of other terrestrial planets can occur during the accumulation of the terrestrial planets, thus altering their initial compositions. This could have supplied the sulfur required to keep Mercury's core molten. The amount of sulfur, and therefore the thickness of the outer fluid core, depends on the amount of mixing.

The proposed Mercury orbiters have a unique opportunity to decide between competing thermal history models, to place tight constraints on the thickness of the fluid outer core, and to estimate the time of onset of solid inner core formation. If the present thickness of the fluid outer core is known, then the amount of core sulfur can be accurately estimated. This, in turn, will constrain the amount of planetesimal mixing from the feeding zones of other terrestrial planets and will provide a much better understanding of terrestrial planet formation. This problem can be addressed by studying Mercury's unique tectonic framework.

The tectonic framework of Mercury consists of thrust faults (lobate scarps) that are more-or-less randomly distributed over the surface viewed by Mariner-10. They may have a global distribution. This system of thrust faults is probably the result of crustal shortening due to cooling of the lithosphere and core. By knowing the lengths and heights of these faults, and based on estimates of the inclinations of their fault planes, it is possible to estimate rather accurately the amount of crustal shortening and, therefore, radius decrease. Furthermore, their age relative to other geologic terranes provides information on the onset of global compression in Mercury's geologic evolution. Estimates of scarp heights and lengths over about 25 percent of the surface viewed by Mariner-10 and extrapolated to the entire planet suggest a radius decrease of about 2 km, a fluid outer core thickness of about 900 km, and an onset of global compression postdating intercrater plains formation. Because of the poor resolution and coverage, however, these estimates are extremely uncertain.

The coverage and resolution of the Mercury orbiter mission will be sufficient to provide the information on the tectonic framework to answer these questions. The combined coverage and resolution by both spacecraft will provide global coverage at better-than-1-km resolution, and about 90 percent coverage at better-than-500-m resolution. It should also result in a substantial amount of stereoscopic coverage. This, combined with shadow measurements, should provide accurate scarp heights. From these data, accurate estimates of amount and onset of global crustal shortening will be possible.

Origin and Composition of Surface Units

One of the outstanding problems of Hermean geology is the origin and extent of plains units. This problem results from the poor coverage and resolution of the Mariner-10 images and the complete lack of compositional information. Currently, both the younger smooth plains and the older intercrater plains are thought to be volcanic, but this interpretation is uncertain, especially for the intercrater plains. The alternate explanation is that they are impact basin ejecta deposits. The mode of formation of these plains units has profound implications for the thermal history of Mercury. If these deposits are volcanic, then during the period of heavy bombardment, Mercury experienced an intense period of volcanic activity which gradually died out and became concentrated in and around large impact basins as compressive stress caused by internal cooling closed off magma sources. This implies that Mercury formed at moderate temperatures, heated up, expanded causing fracturing in a thin lithosphere and extrusion of intercrater plains, and then cooled causing crustal compression. If the deposits are impact basin ejecta, then there was little or no volcanism. This would imply that Mercury formed in a molten state and continuously cooled from that point on, placing the lithosphere in compression and preventing magma from reaching the surface.

The MeO imaging and geochemistry investigations should provide the data to answer this fundamental question. The global coverage and high resolution images (about 70 percent coverage at better-than-400-m resolution) will determine the extent and stratigraphic relationship of these plains units and will allow the discrimination of small-scale structures, such as volcanic domes, flow fronts and sinuous rills, required to determine their origins. Composition variations among these major units determined by the geochemistry experiment will further constrain their origins, and will provide information on the composition of their source regions and the extent and mode of magmatic differentiation.

Impact Processes

The Mercury orbiter global coverage and high-resolution imaging, together with compositional information and gravity data, will permit detailed studies of crater and basin structure, morphology and composition of basin ejecta deposits to reconstruct pre-impact target composition and structure, and post-impact ejecta deposition and modification. This will greatly improve crater statistics and their spatial variations to accurately reconstruct Mercury's geologic and cratering history. High-resolution images of fresh crater morphology, unavailable from Mariner-10 data, will allow detailed comparisons between crater morphologies on the Moon, Mercury, Mars and Venus to better understand how impact parameters, e.g., gravity, impact velocity, and target characteristics affect differences in crater morphology.

Lithospheric Structure

The local gravity field derived from Doppler tracking, combined with imaging and topographic data, will determine whether gravity anomalies are associated with topography and geologic structures. These data can be used to derive variations in lithospheric thickness and to determine whether topography is isostatically compensated. The local gravity field will also determine whether mascons exist and are associated with impact basin fill as they are on the Moon. The Caloris Basin floor has a unique structure indicating that it subsided and then was uplifted. The cause of these tectonic movements is unknown. The determination of the gravity field of this impact basin will be used to derive the internal structure beneath its floor and will greatly aid in interpreting the cause of these movements. These data may be used to derive the thickness of the basin fill.

In summary, the imaging and geochemistry investigations, combined with a knowledge of the local gravity field derived from Doppler tracking, will significantly enhance our current understanding of Mercury. These data should provide the information to address the most fundamental questions concerning Mercury and Solar System science, including the origin of Mercury, and the implications for terrestrial planet formation, and Mercury's thermal history, internal constitution, crustal dynamics, magmatic processes and history, impact processes, and geological and geophysical history.

VI. REFERENCES

1. Report of the Terrestrial Bodies Science Working Group, Volume II. Mercury, JPL Pub. 77-51, Pasadena, California, September 15, 1977.

2. Strategy for Exploration of the Inner Planets: 1977-1987, National Academy of Sciences, Washington, D.C., 1978.

3. Mercury Polar Orbiter, A Proposal to the European Space Agency, July 10, 1985.

4. An Implementation Plan for Priorities in Solar System Space Physics, National Academy of Sciences, Washington, D.C., 1985.

5. Space Science in the Twenty-First Century: Planetary and Lunar Exploration, National Academy of Sciences, 1988.

6. Mercury Dual Orbiter: Mission and Flight System Definition, JPL D-7443, Pasadena, California, 1990.

Appendix A

Spacecraft System Design Team

David H. Collins	–	Team Management and Systems Engineering
Barrie Gauthier	–	Deputy Team Lead and Payload Engineering
Murray A. Koerner	–	Telecommunications Engineering
David P. McGee	–	Power/Pyrotechnics Engineering
Albert S. Yen	–	Attitude and Articulation Control Engineering
Karl Strauss	–	Computer Engineering
David F. Braun	–	Structure/Cabling/Devices Engineering
Carl S. Guernsey	–	Propulsion Engineering
Glenn T. Tsuyuki	–	Thermal Control Engineering

NOTES:

1. Carl W. Buck provided the structure/configuration engineering in FY88 that served as a starting point for work in FY89.

2. Yahya Rahmat-Samii provided an alternative high-gain antenna conceptual design and performance estimate.

3. Burton Zeldin provided solar panel temperature calculations, umbra/penumbra duration calculations, and other data.

4. Funding constraints limited <u>total</u> team effort to $\approx$ 1.5 workweek/week.

www.ingramcontent.com/pod-product-compliance
Lightning Source LLC
Chambersburg PA
CBHW081726170526
45167CB00009B/3721